Poultry Environment Problems

A guide to solutions

Moulton College

NORTHAMPTONSHIRE

Profit through Skill

POULTRY ENVIRONMENT PROBLEMS

A guide to solutions

Editors

D R Charles

A W Walker

Nottingham University Press
Manor Farm, Main Street, Thrumpton
Nottingham, NG11 0AX, United Kingdom

NOTTINGHAM

First published 2002

British Library Cataloguing in Publication Data
Poultry Environment Problems: A Guide to Solutions

ISBN 1-897676-97-2

Disclaimer

Every reasonable effort has been made to ensure that the material in this book is true, correct, complete and appropriate at the time of writing. Nevertheless the publishers, the editors and the authors do not accept responsibility for any omission or error, or for any injury, damage, loss or financial consequences arising from the use of the book.

Typeset by Nottingham University Press, Nottingham
Printed and bound by The Cromwell Press, Trowbridge, Wiltshire

CONTENTS

INTRODUCTION

The analysis and provision of the appropriate climatic environment for poultry has often been debated in recent years. Contention has arisen firstly because the ingenuity of the practical designers of systems has sometimes offered users a bewildering amount of choice. Secondly it has been difficult for users to keep up with the rapidity of the advance of the scientific information.

Despite this the subject is underpinned by a clear guiding rationale. One of the purposes of practical housing and equipment is to provide for the biological, economic and welfare needs of the birds. Fortunately the scientific literature now contains a great deal of information on these needs, so that a recommended approach to any practical housing and ventilation problem is to start with an appraisal of the requirements and responses of the birds. Then, after deciding what the birds require, the designers, engineers and builders can be asked to provide for those needs at a sensible cost, and these days electronic engineers are also often involved. Monitoring is worthwhile, in order to check the ability of the practical equipment to achieve the target environment within sensible tolerances under day to day operating conditions. Compliance with regulations and codes of practice must be checked.

This book offers reviews of the literature on the needs and responses of the birds to temperature, light and air quality. Bird comfort and economic requirements are covered, but many aspects of welfare are not, because there are other books on that subject, not because they are unimportant. Problems are identified and some recommended environmental conditions are concluded from the literature reviewed. The recommendations are tabulated or summarised as appropriate.

The principles of the movement of air through and within buildings, and the characteristics of some typical ventilation systems, are discussed. However, design details of practical commercial housing, equipment and ventilation systems are not covered since they change so quickly, and since there are numerous companies offering them.

The topics chosen for review include those which have been topical in consultancy work over the past few years. Many of the principles discussed apply both to indoor systems of production and to the housing attached to free range systems.

Work on some of these reviews, including the identification and prioritisation of topics, began in 1996, at that time under the co-authorship of Sue Tucker. Tragically she died suddenly in September 1998 at the age of only 37. She was one of those rare individuals with a sound grasp of science combined with an aptitude for practical commercial skills and management. At the time of her death she was running both the research programme at ADAS Gleadthorpe and the ADAS poultry consultancy team. This book owes much to her vision and is dedicated to her memory.

1

RESPONSES TO THE THERMAL ENVIRONMENT

D.R. Charles

The problems

This chapter addresses the following problems associated with the choice of appropriate target thermal environments for several classes of poultry:

- Temperature as affecting metabolism and comfort
- The effects of humidity and air speed on thermal relations
- Quantifying the effects of temperature on feed conversion and therefore on the economics of production. The quantification is affected by:
  - factors such as the effects of temperature on growth rate, egg production, egg weight and feed intake
  - economic optima are not fixed but depend upon prices of products and of feed

Genetic background of the chicken

Gallus gallus, the red jungle fowl, was probably the main progenitor of the domestic fowl (Stevens, 1991), though the Asiatic and Mediterranean breeds may have had different origins. A multi-species origin has often been suggested, though recently Blench and MacDonald (2000) reviewed genetic evidence of Fumihoto *et al.* (1994) refuting it.

Thus there is uncertainty, but it is probable that the chicken's ancestors were from south east Asia, in the warm region approximating to modern Thailand. Interestingly West and Zhou (1989) observed that the natural range of the jungle fowl is bounded by the 10°C January temperature line, so perhaps we should not have been so surprised to discover in the 1960s and 1970s that the optimum temperature of the adult fowl is quite high (see below). In fairness to the prevailing opinion before the 1960s, however, many modern breeds have had a long genetic history of evolution and selection in cooler European climates.

Calorimetry and thermal neutrality

Mount (1974) defined a thermoneutral environment for animals as the range of environmental conditions within which metabolic rate is minimal. Measures of metabolic rate include heat production, oxygen consumption and carbon dioxide output.

Monteith (1974) specified the principles of thermal physiology for animals, and Clark and McArthur (1994) reviewed the physics and physiology of the thermal exchanges of animals.

Calorimetric principles of the energy metabolism of poultry have often been used to estimate a suitable thermal environment. Calorimetric measurements of the metabolic rate of poultry have included those of Romijn and Lokhorst (1966), Balnave (1974), van Kampen (1974), Richards (1974) and Meltzer *et al.* (1982). Reviews on metabolism have included those of Sykes (1977), van Kampen (1981) and Greninger *et al.* (1982). It is interesting that most of these calorimetric estimates found that minimal energy expenditure occurred at environmental temperatures above 20°C, and sometimes much above. This was substantially higher than the practical target house temperatures used at the time. However short term calorimetric measurements do not necessarily simulate long-term practical needs, because of acclimatisation. It is therefore important to review experimental work in which poultry have been maintained at a range of temperatures under long term practical conditions, and such experiments are reviewed below in the sections entitled *Response to temperature*.

An assumption is often made that the comfort zone corresponds with the thermoneutral zone. Although this seems very likely, on the grounds of the survival value of metabolisable energy conservation, there is little firm evidence for it.

The partition of metabolisable energy and the prediction of voluntary feed intake

The laying hen provides an interesting example of the principles of the fate of feed energy in poultry. Metabolisable energy (ME) intake is partitioned three ways in the laying hen. It supplies the energy content of egg output, that of body weight gain and that of heat loss. Of these by far the largest component is heat loss. Emmans (1974) quantified the partition for laying hens and expressed it in the following relationship, which estimates *ad libitum* (voluntary) ME intake:

$$M = bWk + 2E + 5\Delta W$$

where M = voluntary ME intake, (at that time expressed as kcals/ bird day)

W = liveweight, kg
E = egg output, g/bird day
ΔW = weight change, g/bird day
k = a strain specific constant varying from 0.75 to 1
b = a constant depending on strain, temperature and feather quality

Later Emmans and Charles (1977) published values for the constants in the following version of the equation.

$$M = W(a + bT)f + 2E + 5\Delta W$$

Where a had the values 170, 155 and 140, and b had the values -2.2, -2.1 and -2.0, for white, tinted and brown egg laying stocks repectively. The term f was a multiplier for the effect of feathering on heat loss and therefore on the maintenance requirement term W(a + bT).

Table 1.1 VALUES OF THE CONSTANT f FOR THE EFFECT OF FEATHER COVER QUALITY ON MAINTENANCE (EMMANS AND CHARLES, 1977)

Feather loss score	*f*
1	0.94
2	1.00
3	1.08
4	1.20
5	1.40

Subjective feather loss score 1 indicated perfect feathering and score 5 indicated complete feather loss. In practice at that time the plumage condition often changed from about score 2 to score 3 during lay.

The literature contains many other equations for estimating the voluntary ME intake of layers, but that of Emmans and Charles takes into account feathering.

Byerly (1979) tabulated 11 equations, the earliest of which was his own of 1941. Larbier and Leclercq (1994) quoted three more. The various equations give slightly different predictions of course but several of them indicate the maintenance component of requirement increasing as temperature falls.

Of the metabolisable energy (ME) offered to laying hens roughly 2/3 is used as body heat loss and only 1/3 as the energy content of egg output, even under ideal conditions. Example data from a flock at ADAS Gleadthorpe eating 1381 kJ/day per bird illustrates the partition of metabolisable energy (Charles and Alvey, 1995).

Table 1.2 THE PARTITION OF METABOLISABLE ENERGY USED BY THE LAYING HEN, kJ/DAY PER BIRD

Egg energy	459
Liveweight gain	21
Heat loss	901
Total	1381

Note that this partition was for a flock kept at 21°C and with good feather cover. Under conditions of poor house temperature control and poor feathering the proportion of the ME lost as heat would be greater. Wathes and Clark (1981 a and b) described the physics of heat loss from poultry in detail.

Response to temperature - laying hens

Until the 1960s it had been thought that although feed could be saved by keeping layers warm, egg production would be reduced at temperatures above about 15°C. However Bray and Gessel (1961), Payne (1966), Mowbray and Sykes (1971) and Davis *et al.* (1972) realised that provided that the intake of essential non-energy nutrients was maintained, by reformulating the diet to allow for the lower feed intake, then egg output was not depressed up to much higher temperatures, even as high as 30°C; an idea which became known as Payne's hypothesis.

This led to several years of research activity at a number of centres, and ultimately to practical regimes being confidently recommended. Work on large

External and internal views of a facility for applied research at ADAS Gleadthorpe in 1999. The building is an example of the insulated, ventilated, light and temperature controlled housing used throughout the developed world in order to meet the climatic needs of the birds.

numbers of birds in replicated climate rooms contributed substantially to the final acceptance of the use of higher temperatures in the UK, (e.g. Emmans and Charles, 1977). Marsden *et al.* (1987) confirmed Payne's hypothesis over the temperature range 15 to 27°C but found that at 30°C egg output of white and brown strains was depressed whatever diet was fed. Marsden and Morris (1987) quantitatively reviewed 30 published experiments and fitted response equations. They produced some evidence for the existence of a thermo neutral zone at around 20 to 25°C.

It is interesting that it took until the 1970s for standard recommendations for laying house temperature to move up to about 21°C, yet it had been known for many years that metabolic heat output is lowest at about 20°C, (e.g.Hill, 1951, quoting Terroine, 1927).

For the next 20 years or so efforts were made in the UK to save feed by controlling laying house temperature. Practical application included monitoring house temperature with data loggers and recommending simple but effective refurbishments to ventilation systems. The methodology for both the application, and for calculating the cost benefits, was based on techniques such as those published by Spencer (1975), Charles (1981), Charles (1984), Sutcliffe *et al.*(1987), Hill *et al.*, (1988), and Chwalibog and Baldwin (1995). There is more information on monitoring in Chapter 2.

In order that practical recommendations may be made on the same basis as the published experiments it is important to note that in the experiments, by convention, temperatures were usually measured at the feed trough level in the gangways. Within the cages the temperature was often found to be about 3°C warmer, but the difference depended on cage design, stocking density and air change rate. This convention posed difficulties in suggesting where temperature should be recorded in alternative systems such as percheries and on free range. It has become conventional in these systems to measure temperature just above bird height, (see also Chapter 2).

Yahav *et al.* (2000) found no evidence for an effect of relative humidity (rh) on production performance factors of white layers at high temperature (35°C), over a range of regimes from 40 to 45% through to 70 to 75% rh, though body weight was affected. Such a lack of effect is inconsistent with expectations based on the psychrometrics of respiratory heat loss (Chapter 4) and should be interpreted with caution. Presumably the birds were not panting.

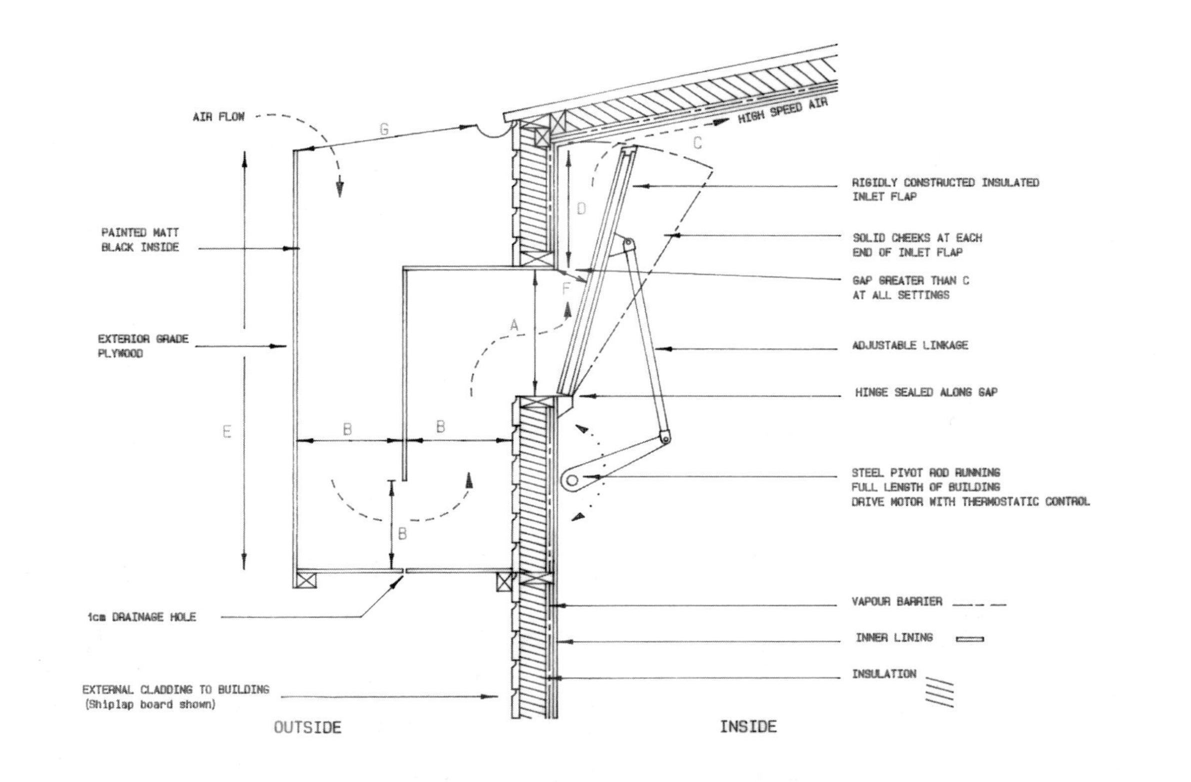

Simple ventilation inlet refurbishments of this kind, with protection from wind, were commonly undertaken in the UK from the 1970s onwards. This was because response experiments had shown that house temperature control needed to be both more accurate and more precise.

Response to temperature - broilers

A large proportion of the ME consumed by a growing animal is lost as heat, (Charles, 1994), and the thermal environment can influence the rate of heat loss. Dry bulb air temperature is the simplest descriptor of the thermal environment, but for broiler chickens and young chicks it is inadequate. Air speed, radiant heat and relative humidity are all capable of modifying the effect of temperature enough to make practical differences.

BROODING

At day old 31°C is probably suitable in draught proof houses, reduced to the finishing temperature at 17 to 21 days. The requirement for dry bulb air temperature is profoundly modified by air speed and by the amount of radiant heat delivered by the brooder. In practice this means that during brooding the supply of ventilation air without draughts is particularly important. Charles (1986) reviewed the literature in more detail, since when Alsam and Wathes (1991 a and b) have greatly refined the information on the brooding environment in systems with a radiant component in which the chicks were able to choose their environment. Stocking density is also likely to modify temperature requirement, through effects on convection from birds and on radiant transfer between birds. Charles (1986) pointed out that the behaviour of the birds could be regarded as a useful indicator of the suitability of a brooding environment and Alsam and Wathes (1991b) considered that "chicks can be more reliable than thermometers". Complications mentioned in the later publications on brooding, such as that of Deaton *et al.* (1996), include possible interactions with ascites.

GROWING

During the growing phase as temperature is increased feed conversion is improved but growth rate is reduced. Thus early work in the ADAS Gleadthorpe climate rooms concentrated on clarifying the balance between these two effects and the consequent effects on gross margin, (Charles *et al.*, 1981; Wathes *et al.*, 1981). At price sets prevailing in the UK during the years following this work the economic optimum was usually within the range 20 to 23°C, depending on age and sex.

Modern broilers grow much faster than the birds used in those experiments. Their rate of metabolic heat production is likely to be higher and we might expect the optimum finishing temperature to gradually fall over the years. Growth is exothermic since the utilisation of ME is at an efficiency of less than 1. Literature reviewed by Larbier and Leclercq (1994) suggested that the energy requirement for live weight gain was 8.8 to 13.0 kJ/g of gain, at an efficiency of 0.58 to 0.85 with a mean value of 0.65. Thus it is possible that as temperature is increased the depression of growth rate sets in at lower temperatures than it did in the early experiments. This possibility is supported by evidence from Homidan *et al.*, (1996), who found a significantly lower rate of weight gain to 49 days at 21°C compared with 19°C.

However in the work of Cheng *et al.* (1997 a and b), using large numbers of plots, though small numbers of birds, the curvatures of responses resembled those reported by Wathes *et al.* (1981). Yahav *et al.* (1996) found a progressive decline in weight gain and feed intake of male broilers from 5 to 8 weeks old as temperature was increased from 18 to 35°C. They fitted quadratic response curves for growth and feed intake against temperature, resembling those of Wathes *et al.* (1981). Maximum growth rate occurred at about 18°C. They found the effects of diurnally cycling temperature regimes difficult to predict, but in general weight gain and feed intake were lower on cycling regimes of large amplitude than in the corresponding average temperature. Rose and Salah Uddin (1997) found similar growth rates at 15, 20 and 25° C but at 30° growth was depressed. Koh and Macleod (1999) recorded faster growth rates at 17°C than at 14° or at temperatures above 17°. The effect was curvilinear from 14 to 32°C, though the observations were made on very small numbers of birds. Howlider (1999) noted that the growth rate of 21 day old broilers declined linearly with increasing temperature from 20.3 to 28.8°C.

Koh and Macleod found that the availability of TME was not much affected by temperature over the range 17 to 32°C.

Response to temperature - broiler breeders

House temperature is likely to affect the amount of feed energy which should be offered to the birds and the published feeding recommendations of authorities such as Leeson and Summers (1997) incorporate this concept. They estimated, for example, that maintenance might take 140 g of peak feed requirement/day per bird at 18°C, but only 125 at 24°C. Corresponding values for total feed were 180 and 165 respectively.

The equation of Emmans and Charles (1977) has been used to make the following estimates of feed intakes which might occur on *ad libitum* feeding for a feed providing 11.55 MJ/kg ME, and assuming that breeders have the same values of a and b as brown commercial layers. However, note that there are no experimental grounds for this assumption, so that the values in Table 1.3 are comparisons of possible temperature effects: they are not recommendations. The estimates are for birds of 3.4 kg liveweight at 30 weeks of age, gaining weight at 4 g/bird per day, at 83.6% rate of lay, and with an egg weight of 60 g/egg.

Table 1.3 ESTIMATES OF THE EFFECT OF TEMPERATURE ON *AD LIBITUM* FEED INTAKE OF BROILER BREEDER FEMALES

Feather score	*House temperature, °C*	*Feed intake, g/bird day*
2	15	179
2	20	167
2	25	155
3	15	190
3	20	177
3	25	163

An interaction between house temperature, feather score and ME allowance seems almost inevitable and there may be an optimum temperature for performance which is sometimes not being achieved in winter in UK at the time of writing. There is therefore scope for research on the energy balance of broiler breeders and for practical evaluations of the economics of heating in winter. Broiler breeders are not normally fed *ad libitum* and in practice under conditions of varying temperature it may be difficult to judge the appropriate feed allowance day by day. Individual variation complicates practical feeding further. It may be worth considering the examination of techniques such as automatic monitoring of bird weight.

At present it is not possible to confidently state an optimum temperature for performance in the absence of sufficient data from suitable response experiments. The recommendations of the breeding companies for the energy balance of their birds should be carefully followed.

Response to temperature - growing turkeys

Some of the experiments of the 1960s and 1970s on the temperature requirements of growing turkeys used replicated pens in paired but unreplicated buildings, and were therefore not fully capable of distinguishing temperature effects from building effects. Charles (1989) reviewed the work up to that time, including replicated and unreplicated work, and including some data from replicated climate rooms in the early 1980s at ADAS Gleadthorpe. As might be expected *ad libitum* feed intake was generally found to decline with increasing temperature, and the response of growth rate to temperature was sometimes reported to be quadratic, suggesting the existence of an optimum, though in work reviewed on males of stocks available in the early 1980s, and grown to 24 weeks of age, both growth rate and feed intake declined over the temperature range from 13° to 23°C.

Modelling techniques described by Charles (1989) were used at that time to permit economic optima to be calculated, allowing for the money values of feed and of liveweight. The higher the value of the liveweight the lower the optimum temperature. For example, birds for the fresh Christmas market could be grown at relatively low temperatures while those for the lower value frozen carcass market needed the better feed conversions associated with slightly higher house temperatures. Thus, due to the complications of differing markets and differing ages at slaughter there is no single economic optimum temperature for growing turkeys.

Summary of recommendations

Table 1.4 SUMMARY OF TEMPERATURE RECOMMENDATIONS

Class of stock	*Suggested temperatures, °C*	*Factors affecting*
Laying hens	19 to 22	*Relative money values of egg numbers, egg weight and feed intake. Feathering. Type of production system may affect optimum. Published values are from cages, measured in the gangways.

Table 1.4 (Contd)

Broilers, brooding (draught free houses)	31 at day old, reducing to finishing temperature at 17 to 21 days	Air speed, radiative environment, stocking density, age, breed and sex all have effects. Use chick behaviour as a guide. Chicks should move freely in and out of clusters, neither persistently huddling nor persistently avoiding heat.
Broilers, growing	18 to 22	*Relative money values of liveweight and feed
Broiler breeder females	Not certain	Affected by feathering
Growing turkeys	Perhaps 12 to 20 based on published data reviewed, but data on modern stocks are needed	Relative money values of liveweight and feed

*Note that the estimated monetary effects of unit deviation from the optimum temperature may increase curvilinearly with increasing deviation.

References

Alsam, H. and Wathes, C.M. (1991a) Conjoint preferences of chicks for heat and light intensity. *British Poultry Science* **32:** 899-916

Alsam, H. and Wathes, C.M. (1991b) Thermal preferences of chicks brooded at different air temperatures. *British Poultry Science* **32:** 917-927

Balnave, D. (1974) Biological factors affecting energy expenditure. In: *Energy requirements of poultry.* Edit. Morris, T.R. and Freeman, B.M., British Poultry Science Ltd., 25-46

Blench, R. and MacDonald, K.C. (2000) Chickens. In: *The Cambridge world history of food.* Edit. Kiple, K.F. and Ornelas, K.C., Cambridge University Press, 496-498

Bray, D.J. and Gessel, J.A. (1961) Studies with corn soya diets. 4. Environmental temperature - a factor affecting performance of pullets fed diets sub-optimal in protein. *Poultry Science* **40:** 1328-1335

Byerly, T.C. (1979) Prediction of the food intake of laying hens. In: *Food intake regulation in poultry.* Edit. Boorman, K.N. and Freeman, B.M., British Poultry Science Ltd., Edinburgh, 327-363

Charles, D.R. (1981) Practical ventilation and temperature control for poultry. In: *Environmental aspects of housing for animal production.* Edit. Clark,J.A. Butterworths, London, 183-196

Charles, D.R. (1984) A model of egg production. *British Poultry Science* **25:** 309-322

Charles, D.R. (1986) Temperature for broilers. *World's Poultry Science Journal* **43:** 249-258

Charles, D.R. (1989) Environmental responses of growing turkeys. In: *Recent advances in turkey science*. Edit. Nixey, C. and Grey, T.C., Butterworths, London, 201-214

Charles, D.R. (1994) Comparative climatic requirements. In: *Livestock housing*. Edit. Wathes, C.M. and Charles, D.R., CAB International, Wallingford, 3-24

Charles, D.R. and Alvey, D.A. (1995) Feathering and food conversion in layers. *ADAS Poultry Progress* No. 22, May 1995

Charles, D.R., Groom, C.M. and Bray, T.S. (1981) The effects of temperature on broilers: interactions between temperature and feeding regime. *British Poultry Science* **22:** 475-482

Cheng, T.K., Hamre, M.L. and Coon, C.R. (1997a) Effect of environmental temperature, dietary protein, and energy levels on broiler performance. *Journal of Applied Poultry Science Research* **6:** 1-17

Cheng, T.K., Hamre, M.L. and Coon, C.R. (1997b) Responses of broilers to dietary protein levels and amino acid supplementation to low protein diets at various environmental temperatures. *Journal of Applied Poultry Research* **6:** 18-33

Chwalibog, A. and Baldwin, R.L. (1995) Systems to predict the energy and protein requirements of laying fowl. *World's Poultry Science Journal* **51:** 187- 196

Clark, J.A. and McArthur, A.J. (1994) Thermal exchanges. In: *Livestock housing*. Edit. Wathes, C.M. and Charles, D.R., CAB International, Wallingford, 97-122

Davis, R.H., Hassan, O.E.M. and Sykes, A.H. (1972) The adaption of energy utilisation in the laying hen to warm and cool ambient temperatures. *Journal of Agricultural Science, Cambridge* **79:** 363-369

Deaton, J.W., Branton, S.L., Simmons, J.D. and Lott, B.D. (1996) The effect of brooding temperature on broiler performance. *Poultry Science* **75:** 1217-1220

Emmans, G.C. (1974) The effects of temperature on the performance of laying hens. In: *Energy requirements of poultry*. Edit. Morris, T.R. and Freeman, B.M. British Poultry Science Ltd., Edinburgh, 79-90

Emmans, G.C. and Charles, D.R. (1977) Climatic environment and poultry feeding in practice. In: *Nutrition and the climatic environment*, Edit. Haresign,W., Swan,H. and Lewis,D., Butterworths, London, 31-50

Greninger, T.J., DeShazer, J.A. and Gleaves, E.W. (1982) Simulation model of poultry energetics for developing environmental recommendations. In: *Livestock environment. II.* American Society of Agricultural Engineers, 234-240

Hill, S.R. (1951) A consideration of the problems involved in ventilating the poultry laying house. *Poultry Science* **30:** 558-568

Hill, J.A., Charles,D.R., Spechter, H.H., Bailey, R.A. and Ballantyne, A.J. (1988) Effects of multiple environmental and nutritional factors on laying hens. *British Poultry Science* **29:** 499-512

Homidan, A. al, Robertson, J.F. and Petchey, A.M. (1996) Some factors affecting dust and ammonia production in broiler houses. *World's Poultry Science Association (UK Branch), Proceedings of spring meeting*, Scarborough, 118

Howlider, M.A.R. (1999) Interaction of temperature and stocking density on growth, meat yield and meat quality of broilers. *Symposium on the quality of poultry meat.* WPSA, Bologna, 277-282

Koh, K. and Macleod, M. (1999) Effects of ambient temperature on heat increment of feeding and energy retention in growing broilers maintained at different food intakes. *British Poultry Science* **40:** 511-516

Larbier, M. and Leclercq, B. (1994) *Nutrition and feeding of poultry.* Translated and edited Wiseman, J., Nottingham University Press and INRA, Loughborough

Leeson, S. and Summers, J.D. (1997) *Commercial poultry nutrition.* University Books Guelph, Ontario

Marsden, A. and Morris, T.R. (1987) Quantitative review of the effects of environmental temperature on food intake, egg output and energy balance in laying pullets. *British Poultry Science* **28:** 693-704

Marsden, A., Morris, T.R. and Cromarty, A.S. (1987) Effects of constant environmental temperatures on the performance of laying pullets. *British Poultry Science* **28:** 361-380

Meltzer, A., Goodman, G. and Fistool, J. (1982) Thermoneutral zone and resting metabolic rate of growing white leghorn-type chickens. *British Poultry Science* 23: 383-391

Monteith, J.L. (1974) Specification of the environment for thermal physiology. In: *Heat loss in animals and man.* Edit. Monteith, J.L. and Mount, L.E., Butterworths, London, 1-17

Mount, L.E. (1974) The concept of thermal neutrality. In: *Heat loss in animals and man.* Edit. Monteith, J.L. and Mount, L.E., Butterworths, London, 425-439

Mowbray, R.M. and Sykes, A.H. (1971) Egg production in warm environmental temperatures. *British Poultry Science* **12:** 25-29

Payne, C.G. (1966) Environmental temperature and egg production. In: *The physiology of the domestic fowl*, Edit. Horton-Smith,C. and Amoroso, E.C., Oliver and Boyd, Edinburgh, 235-241

Richards, S.A. (1974) Aspects of physical thermoregulation in the fowl. In: *Heat loss in animals and man*. Edit. Monteith, J.L. and Mount, L.E., Butterworths, London, 255-275

Romijn, C. and Lokhorst, W. (1966) Heat regulation and energy metabolism in the domestic fowl. In: *Physiology of the domestic fowl*. Edit. Horton-Smith, C. and Amoroso, E.C., Oliver and Boyd, Edinburgh, 211-227

Rose, S.P. and Salah Uddin, M. (1997) The effect of temperature on the response of broiler chickens to lysine balance in the dietary crude protein. *World's Poultry Science Association (UK Branch) Proceedings of spring meeting,* Scarborough, 29-30

Spencer, P.G. (1975) A cost benefit analysis of temperature controls in housing laying hens. *World's Poultry Science Journal* **31:** 309

Stevens, L. (1991) *Genetics and evolution of the domestic fowl*. Cambridge University Press, Cambridge

Sutcliffe, N.A., King, A.W.M. and Charles,D.R. (1987) Monitoring poultry house environment. In: *Computer applications in agricultural environments*. Edit. Clark,J.A., Gregson, K. and Saffell, R.A., Butterworths, London, 207-218

Sykes, A.H. (1977) Nutrition-environment interactions in poultry. In: *Nutrition and the climatic environment*. Edit. Haresign, W., Swan, H, and Lewis, D., Butterworths, London, 17-29

Terroine, E.F. and Trautman, S. (1927) Influence de la temperature exterieure sur la production calorique des homeothermes et lois des surfaces. *Annals of Physiology* **31:** 422-457

van Kampen, M. (1974) Physical factors affecting energy expenditure. In: *Energy requirements of poultry*. Edit. Morris, T.R. and Freeman, B.M., British Poultry Science Ltd., Edinburgh, 47-59

van Kampen, M. (1981) Thermal influences on poultry. In: *Environmental aspects of housing for animal production*. Edit. Clark, J.A., Butterworths, London, 131-147

Wathes, C.M. and Clark, J.A. (1981a) Sensible heat transfer in the fowl: boundary layer resistance of a model fowl. *British Poultry Science* **22:** 161-173

Wathes, C.M. and Clark, J.A. (1981b) Sensible heat transfer in the fowl: thermal resistance of the pelt. *British Poultry Science* **22:** 175-183

Wathes, C.M., Gill, B.D. and Back, H.L. (1981) The effects of temperature on broilers: a simulation model of the responses to temperature. *British Poultry Science* **22:** 483-492

West, B. and Zhou, B-X. (1989) Did chickens go north? New evidence for domestication. *World's Poultry Science Journal* **45:** 205-218

Yahav, S., Shinder, D., Razpakovski, M., Rusal, M. and Bar, A. (2000) Lack of response of laying hens to relative humidity at high ambient temperature. *British Poultry Science* **41:** 660-663

Yahav, S., Straschnov, A., Plavnik, I. and Hurwitz, S. (1996) Effects of diurnally cycling versus constant temperatures on chicken growth and food intake. *British Poultry Science* **37:** 43-54

2

VENTILATION RATE REQUIREMENTS AND PRINCIPLES OF AIR MOVEMENT

D.R.Charles, [1]J.A.Clark and the late S.A.Tucker
[1]University of Nottingham, Sutton Bonington Campus, Loughborough, LE12 5RD, UK

The problems

Designers and users of poultry ventilation systems may need to address the following problems:

- Quantification of the maximum ventilation rate requirement for the elimination of spare metabolic heat
- Quantification of the minimum ventilation rate requirement for the provision of good air quality
- Factors affecting the movement of air through buildings
- Characterisation of ventilation systems

Historical background

The design of ventilation systems for poultry housing has seen a certain amount of development by trial and error, often with good reason in the absence of any other guidance. However, this review is confined to the consideration of scientific principles and standard practices based upon them. The detail of installation is the province of commercial design engineers.

The amount of ventilation air needed by poultry has been contentious ever since intensive housing became popular. It is still contentious. Quoted figures are frequently confused by failure to take into account system losses and resistance to air flow, thus confusing real ventilation rates with gross values of fan capacity, but in addition there are genuine differences of opinion about ventilation requirements. Users are also still occasionally confused by the need to distinguish between the maximum ventilation rate requirement and the minimum.

The importance of air supply in poultry houses has long been taken seriously. Robinson (1948) wrote: "Ventilation of the poultry house is of supreme importance. It is a matter to which considerable thought should be given, because failure to provide an abundance of fresh air without draught is by far the most common cause of respiratory diseases to which poultry are heir."

The following quotation appeared in the poultry press even earlier (Hawk, 1910). "Whatever style of poultry house may be selected, it is necessary to have it well ventilated............unless there is a fair amount of space, and good ventilation, the atmosphere soon becomes unhealthy and even poisonous." Earlier still in 1851 Trotter observed that "...it is therefore of the utmost importance that the ventilation be of the most perfect description." The systematic study of the requirements for air composition seems to date from Mitchell and Kelley (1933), who provided estimates of the heat, moisture and carbon dioxide outputs of poultry. Their estimates are still useful and still quoted. As early as 1926, however, Dann had published estimates of moisture outputs and had realised their consequences for litter management.

Maximum ventilation rate

The necessary capacity of the ventilation system, whether driven by fans or by natural convection, is determined by the maximum ventilation requirement. It was realised remarkably early in the development of the technology of indoor poultry production that the maximum requirement is the amount of air needed to remove metabolic heat during warm weather.

Several authors developed recommendations based on calculations of house heat balance (Cropsey, 1951; Hill, 1951; Esmay, 1958; Sainsbury, 1959; Longhouse *et al.*, 1960 and Payne, 1961; reviewed by Charles, 1970). The intention was to prevent the equilibrium house temperature rising more than about 3°C above outside temperature. It was also recognised that without air conditioning the equilibrium inside temperature cannot be held below outside shade air temperature. The house heat balance equation, expressed in modern S.I. units (Système Internationale d'Unités) as described by Monteith (1984), was published by Saville *et al.* (1978), and its use in practical housing was described by Charles (1981) in the following format.

$$V = (Q_s + Q_h - (UA\ \Delta T))/(h\ \Delta T)$$

where: ΔT = acceptable temperature lift above outside temperature, K
Q_s = sensible heat output, W/bird
Q_h = solar heat penetration, W/bird
V = ventilation rate, (m^3/s) per bird
U = average thermal transmittance of the walls and roof, W/(m^2K)
A = exposed area of walls and roof, m^2/bird, (normally greater than the floor area per bird by a factor of about 1.2 to 1.7, depending on the shape of the building).
h = Volumetric heat capacity of air, J/(m^3K), (approximately 1200 J/(m^3K) at 20°C and normal atmospheric pressure)

It has generally remained conventional to calculate a required maximum ventilation rate, V_{max}, such that ΔT does not exceed 2.5 to 3°C. The reason is diminishing benefit. To achieve slightly smaller differences requires much greater air change rates, and therefore a much more expensive installation. Table 2.1 illustrates the point. Thus traditional recommendations in UK have included, for example, 2.7 (m^3/s) per thousand 2.2 kg layers, and 3.1 (m^3/s) per thousand 2.7 kg broilers (Charles *et al.,* 1994).

Table 2.1 THE EFFECT OF MAXIMUM VENTILATION RATE FOR LAYING HENS ON TEMPERATURE LIFT ABOVE OUTSIDE TEMPERATURE. FOR THESE EXAMPLES Q_s WAS TAKEN AS 8.5 W/BIRD, Q_h AS ZERO, U AS 0.35 W/(m^2K) AND A AS 0.075 m^2/BIRD

V, (m^3/s) per 1000 birds	*ΔT, °C*
1.0	6.9
1.5	4.7
2.0	3.5
2.5	2.8
3.0	2.3
3.5	2.0
4.0	1.8
5.0	1.4
9.0	0.8

All calculations of the ventilation rate provided, when choosing fans, should allow for system working pressure, which is normally at least 50 Pa. It is also important to point out that a ventilation rate deemed adequate for the building may be inadequate at particular locations within the building if the distribution of air is poor.

In many circumstances the solution of the house heat balance equation gives good prediction of actual house temperature. However, there are four sources of imprecision.

Firstly, some terms are not included in the simplified equation quoted above. Strictly, terms could be included for factors such as heat storage in both the building and in the bodies of the birds, floor and edge heat losses and gains, heat losses and gains due to evaporation from litter and manure, heat gains from litter microbial metabolism and heat gains from electric lights and motors. Most of these, except storage, are of small magnitude, so that at equilibrium errors introduced by excluding them are also small. It is assumed for the purposes of calculating maximum ventilation rate that there are no heat inputs from heaters and brooders, since these should be switched off in warm weather.

Secondly, Wilson (1986) found that the assumption that Q_h is zero is frequently incorrect. Solar gains of up to 30 W/m^2 were measured penetrating the roofs of old broiler houses in the Midlands of England in summer. Haywood (1990) found that the solar gain through the roofs of single skin cattle buildings in Britain could be as much as 85 W/m^2 in clear summer weather. Q_h will be low in buildings with a high standard of insulation, maintained in good condition, and where the roof reflectivity is high, but it is probably seldom zero. This also applies to the wall heat flux.

Thirdly, Table 2.1 uses a constant value, 8.5 W/bird, for Q_s . In fact, Q_s has been shown to vary with temperature, stocking density, time of day and feather cover, as well as with liveweight. Table 2.2 summarises some of the literature on sensible heat loss. At high temperature, evaporative heat loss becomes significant if panting occurs, and this results in reduced values for Q_s (Richards, 1974).

Fourthly, it is clear from Table 2.2 that Q_s is not a linear function of liveweight. Therefore neither is V. A broiler chick weighing 0.05 kg produces 14 W/kg whereas a 2 kg bird produces only 2.7 W/kg. Therefore it became customary to express the required ventilation rate per kg$^{0.75}$ following the style of the metabolic work of, for example, Brody (1945) and Kleiber (1961). Thus, Charles (1994) suggested that the maximum ventilation requirement for a bird of any weight could be expressed as 1.5 x 10^{-3} (m^3/s) for each kg$^{0.75}$ of its liveweight. The value of the coefficient 1.5 was increased to between 1.55 and 1.6 a few years later in response to genetic changes (see below). The use of the exponent

Table 2.2 SUMMARY OF SOME OF THE PUBLISHED MEASUREMENTS OF SENSIBLE HEAT LOSS

Stock	*Liveweight, kg/bird*	*Temperature °C*	Q_s, *W/bird*	*Reference*	*Notes*
Layers	1.6 (white breed)	23	8.7 7.4	Olson *et al.* (1974)	1 per cage 2 per cage
	(brown breed)	15 20 25	9.9 8.8 5.0	Richards (1977)	1 per cage Well feathered
	(brown)	15 20 25	14.9 13.4 9.5	Richards (1977)	1 per cage Poorly feathered
	1.5 (white) 2.2 (brown)	20 20	day 8.0 night 5.0 day 9.7 night 5.8	Lundy *et al.* (1978)	1 per cage
	1.8	20	8.35 (Q_s = 19.91-0.578T), where T=air temperature	von Wachenfelt *et al.* (2001)	Measured in aviaries. Diurnal variation and linear decline with temperature found
Broiler breeders			day 11 night 14	MacLeod *et al.*, (1980)	
Broilers	0.04 0.36 0.95 1.36 2.04	29 25 19 19 19	0.7 3.4 6.7 8.6 9.6	Longhouse *et al.* (1968)	
	1.6 1.5	16 29	6.7 4.7	Reece *et al.* (1969) Deaton *et al.* (1969)	

Table 2.2 (Contd)

Stock	Liveweight, kg/bird	Temperature °C	Q_s, W/bird	Reference	Notes
	0.05	25	0.7	Wathes (1978)	Allowing for clustering in large groups
	0.12	24	1.3		
	0.40	23	3.4		
	2.04	16	5.6		
	3.0	24	11.7	Xin *et al.* (1996)	
Turkeys	0.11	35	0.5	De Shazer *et al.* (1974)	
	0.24	32	1.4		
	0.42	29	2.0		
	0.63	27	3.7		
	0.96	24	5.6		
Ducks	2	7	24.6	Pugh (1978)	
		11	23.4		
		17	13.0		
		22	12.0		
		26	10.6		

0.75 does not perfectly linearise the relationship between weight and Q_s, and is not intended to imply precision. Therefore some authors and editors insist on writing 3/4 rather than 0.75.

Minimum ventilation rate

The early writers on ventilation for poultry housing (e.g. Mitchell and Kelley, 1933; Davies, 1951) realised that the minimum ventilation requirement, used when the conservation of metabolic heat is desirable, is set by different criteria to those described above for the maximum. The criteria for minimum ventilation are essentially determined by the acceptability of air quality, and include the concentrations of ammonia, carbon dioxide, moisture, dust, odours and micro-organisms. The control of moisture content was rightly regarded as important by the early authors. Sainsbury (1959) and Payne (1959) considered that house relative humidity should be kept below 80% during cold weather.

Hill (1951) quoted Emmel (1941), who had observed increased mortality in battery hens in poorly ventilated houses, mainly attributed to avian leukosis complex. Authors such as Harry (1964) and, later, Wathes (1994) developed concepts of air hygiene which are essential to the satisfactory design and operation of ventilation systems.

The evidence reviewed below concerns the effects of air quality on animal performance and welfare. Effects on staff working in the buildings are, of course, also important, but are not within the scope of this review. However in many countries, including the UK, health and safety regulations impose constraints on the exposure of personnel to concentrations of noxious materials. Wathes (1994) quoted the limits set for staff exposure by the UK Health and Safety Executive (1992). These are generally higher than the target levels for poultry suggested below, although workers are generally exposed to contaminants for fewer hours per day than are the birds.

GAS COMPOSITION - CARBON DIOXIDE, OXYGEN AND AMMONIA CONCENTRATIONS

Estimates of the maximum physiologically tolerable concentrations of carbon dioxide have a long history. Mitchell and Kelley (1933) suggested keeping the concentration of CO_2 below 0.5%, but preferably below 0.1%, on the evidence of layer performance. Practical calculations of minimum ventilation rate, particularly those of Davies (1951), were based on this work and on studies of the ventilation needs of humans. Currently, the usual suggestion is that CO_2 concentration should be kept below 0.3%, but the recommendation is not based on any new evidence (e.g. Charles, 1994).

Air change rates adequate for the control of CO_2 have traditionally been regarded as likely to be sufficient for oxygen supply because of the very small drop below atmospheric content of 21%. Current concerns about ascites in broilers have redirected attention towards the possible effects of oxygen depletion, and this seems a good reason to aim for concentrations of CO_2 below 0.3%. Mitchell (1996) pointed out that many of the lesions and physiological alterations observed in ascitic birds are consequences of right ventricular failure and /or chronic hypoxaemia. However birds have an inherently more efficient lung system than mammals. Becker *et al.* (1995a) found a lower rate of weight gain and a higher incidence of ascites in day old chicks at 17.3% oxygen, but such low levels would not occur in practice according to measurements of oxygen consumption by MacLeod (1990). Becker *et al.* (1995b) found that mortality was affected at 13.6% oxygen.

The amount of oxygen depletion is usually close to the amount of carbon dioxide increase, though changes in the respiratory quotient could make small differences. Therefore 0.3% carbon dioxide is likely to be associated with approximately 20.7% oxygen.

In practice, the control of ammonia concentration is often the principal determinant of minimum ventilation rate. It is generally agreed that the concentration should be kept as low as possible, even if this necessitates exceeding tabulated values for the minimum ventilation rate, since several deleterious effects have been documented. Charles (1993) reviewed the literature on the subject, and summarised some landmark findings as in Table 2.3. Wathes (1998) considered, on the basis of a small sample of emission rates, that much higher minimum ventilation rates than those in Tables 2.4 and 2.5 may sometimes be necessary to control ammonia.

Table 2.3 THE EFFECTS OF AMMONIA ON POULTRY

Effect	*Date first documented*	*Author*
Keratoconjunctivitis incidence	1950	Bullis *et al.*
Indicator of poor ventilation	1952	Scorgi and Willis
Adversly affects broiler growth	1964	Valentine
Adversly affects egg production	1964	Charles and Payne
Increases disease risk	1964	Anderson *et al.*
Air sac lesions	1978	Oyetunde *et al.*
Affects septa thickness	1984	Beck and DeShazer

AIR MOISTURE CONTENT

The control of the air moisture content is also an important function of minimum ventilation. There are practical physical considerations, such as the prevention of condensation on surfaces, which are also profoundly affected by the standard of house insulation and its vapour check (Wathes, 1981). In moist temperate climates, such as that of the UK, the recommended standards of insulation, (approximately $0.35 W/(m^2 K)$ or better), are based partly on such considerations.

Another consideration is the influence which air moisture content has on litter condition. Payne (1967) found a correlation between litter quality and the relative humidity (rh) (% of saturation moisture holding capacity) during the previous week. A mean weekly rh above 72% was associated with poor litter. Tucker and Walker (1992) calculated the effects of ventilation rate and insulation on house rh and on expected litter condition and confirmed the standard UK insulation recommendations.

The moisture content of the air is, of course, an extremely important variable during heat stress, (see Chapter 3), but that is beyond the present scope on the grounds that minimum ventilation is inappropriate during heat stress.

The relative humidity can be read from a psychrometric chart (e.g. Clark and McArthur, 1994) if the air temperature and absolute humidity, M (g/m^3), are known. Likewise, if measurements of rh and temperature are available then M may be estimated, as well as the effect of increments in M on the rh and on the dew point temperature of surfaces. Dew point is the temperature at which the air becomes saturated, and is therefore the surface temperature below which condensation takes place.

DUST AND MICRO-ORGANISMS

In a comprehensive review of evidence up to that time Carpenter (1986) found that the dust in livestock buildings is almost entirely organic in origin, and in layers buildings was mainly composed of skin and feather debris. Amounts released varied from 2.5 to 90 mg/hour per bird. Grub *et al.* (1965) identified sources of poultry dust as litter, feed, skin and feathers, and the data of Koon *et al.* (1963) on the protein content of poultry dust suggested that feed was a less important source than it was in pig housing.

The air within animal buildings contains micro-organisms, though there are no generally accepted levels of tolerance of non-pathogenic organisms (Wathes, 1994). Some of the organisms are attached to dust particles (Carpenter, 1986). Particulate behaviour in air is determined by aerodynamic size (Wathes, 1994). Gordon (1963) found indications that the bacterial concentration of air was lowest in piggeries with high air absolute humidity. In his observations, the value of sedimentation rate in reducing the bacterial complement of high temperature, high humidity, piggeries was striking. Later Webster (1981) noted in a review that the sedimentation of air borne pathogens may be accelerated at high rh.

Research reviewed by Webster (1981) suggested that the death rates of *E. coli* and mycoplasmas may be more rapid at 70% rh than in both very dry or very wet air. Rhinoviruses may survive best at high humidities but some viruses survive best at low humidities. Wathes (1994) wrote in a review that viruses with structural lipids are not stable at low rh, while those lacking a lipoprotein envelope survive best at high rh. The response of air borne bacteria to rh was considered to be species dependent.

The balance of evidence suggests that a relative humidity of about 60-65% might be an appropriate recommendation, to avoid problems of disease dissemination at low and high rh, and this is one of the determinants of the minimum ventilation rate.

MASS BALANCE

The concentration of an air contaminant is determined by the balance between inflow and outflow. Wathes (1994) provided a mass balance equation.

COMBINED EFFECTS ON VENTILATION RATE REQUIREMENT

Taking into account all the above factors it was for some years often agreed that the minimum ventilation rate requirement was approximately (1.6 to 2) x 10^{-4} (m^3/s) per $kg^{0.75}$ for poultry of most weights and species (Charles, 1994). This is about 0.29 (m^3/s) per thousand 2.2 kg layers, or 0.34 (m^3/s) per thousand 2.7 kg broilers.

Such values have seldom been tested experimentally because of the high cost of realistic experiments, and for layers response curves derived from work on the effects of graded levels of ventilation rate on performance are not available. However, Emmans and Charles (1977) compared the performance of birds exposed to 0.25 and 0.5 (m^3/s) per 1000 layers in replicated climate rooms. There were no consistent differences in performance and the lower value was taken as sufficient. Charles *et al.* (1981) compared 1.1, 1.5, 3 and 4.5 x 10^{-4} (m^3/s) per $kg^{0.75}$ for broilers at 20°C after brooding in one experiment and 0.4,

1.1, 1.5 and 3 x 10^{-4} (m^3/s) per $kg^{0.75}$ in another. A third experiment used 0.4 and 1.5 x 10^{-4} (m^3/s) per $kg^{0.75}$ at each of 20 and 25°C after brooding. Effects of ventilation rate on performance were not significant in any of the experiments, although on the basis of the rank order of feed conversion efficiencies (1.5 to 1.6) x 10^{-4} (m^3/s) per $kg^{0.75}$ was considered to be sufficient.

In the years since that time there has been considerable genetic change. Due to substantially higher rates of egg output and broiler growth it is likely that modern birds of a given body weight have a higher metabolic rate than their predecessors. Thus it has been suggested, (personal correspondence with Nick Lynn of the Cobb Breeding Co. Ltd., 1997), that the coefficients for the maximum and minimum ventilation rates should be increased. One possible method of calculating the appropriate increase is based upon the assumption that heat production provides a satisfactory index of metabolism, and that therefore factors relevant to ventilation, such as carbon dioxide output, will be proportional to it. An attempt at such an update has been made as follows.

Rates of production in the early 1980s were compared with current rates. The partition of heat production between maintenance and production for layers was derived from the data of Emmans (1974) and from the partition equations of Emmans and Charles (1977). Maintenance heat production per kg of body weight was assumed to be the same for the former and the current birds, while the heat associated with the extra egg output of modern stocks was used to estimate the extra heat produced. Similar procedures were used for broilers using estimates on the partition of heat production quoted by Larbier and Leclercq (1994). As a result the coefficients given in Table 2.4 and the recommendations in Table 2.5 were derived. Growing turkeys were regarded as having similar needs to those of broilers.

Higher rates of ventilation may be necessary where the heaters or brooders deplete oxygen or add to the carbon dioxide in the house air.

Interestingly, while there seems to be general agreement amongst both specialists and the industry that these higher ventilation rates are probably now necessary, there has been one recent publication suggesting that the resting metabolic rate of modern broilers is lower than expected (Gavin *et al.,* 1998). They suggested that this was because most key metabolic organs, such as the intestines, heart and liver, have not increased in size with increasing body weight, thus there is a lower energy cost of maintaining these tissues.

Summary of requirements

Maximum ventilation rate = $a \times 10^{-3}$ m^3/s per $kg^{0.75}$

Minimum ventilation rate = $b \times 10^{-4}$ m^3/s per $kg^{0.75}$

where the values of a and b are:

Table 2.4 VENTILATION COEFFICIENTS

	a	*b*
Layers	1.55	1.6 to 2
Broilers	1.6	1.9 to 2
Growing turkeys	1.6	1.9 to 2

Table 2.5 EXAMPLES OF VENTILATION RATE RECOMMENDATIONS FOR POULTRY

Stock	*Liveweight, kg/bird*	*Requirements* Maximum, (m^3/s) per 1000 birds	*Requirements* Minimum, (m^3/s) per 1000 birds
Pullets and layers	2.0	2.6	0.27 to 0.34
	2.5	3.1	0.32 to 0.40
	3.5	4.0	0.41 to 0.51
Broilers	0.2		0.06 to 0.06
	0.8		0.16 to 0.17
	2.2		0.34 to 0.36
	2.7	3.4	0.40 to 0.42
Turkeys	2	2.7	0.32 to 0.34
	5	5.3	0.64 to 0.67
	10	9.0	1.07 to 1.12

Ventilation control

House temperature control is normally achieved by modulating the ventilation rate between the two limits (maximum and minimum) using a thermostatic device sensing the house temperature. There are many versions of such devices, nowadays usually electronic.

In a practical installation it is important that the thermostatic device is situated in a position likely to represent the environment experienced by the average bird in the building. In the case of cages, temperature recommendations are defined, both by convention and in order to replicate the response experiment data, (see Chapter 1), as gangway temperatures rather than as temperatures within the cages. For floor systems the temperature just above bird height is used. To ensure that their output represents house conditions, sensors should be protected from the flow of cold incoming air, and they should be remote from spot heat sources such as light bulbs. Relevant control theory was discussed by Randall (1981) and by Randall and Boone (1994). Earlier descriptions of the general principles of the use of maximum and minimum ventilation rates were given by Dobson *et al.* (1969), Charles and Spencer (1976) and Charles (1981).

Traditionally the minimum ventilation rate has been manually selected, or built into the control system, but over-ridden if and when ammonia concentrations are unacceptable. There are currently developments in sensors for the control of ventilation, however, which operate by directly sensing the air quality. It would probably be wise to ensure that such systems are not permitted to provide less than the tabulated minimum air change rates, and that great care is taken with their calibration and with the checking of the stability over time of the sensor responses. Built in minima would be wise in order to cover for possible sensor failure. Whatever system of control is chosen it is important that air quality, as defined above, should be acceptable.

Moving air through buildings

There are several forces which drive air through a building. Electric fans provide the main intentional force in most controlled environment systems, but there are systems which depend for air movement upon buoyancy forces and upon wind pressure. These are commonly called natural ventilation systems.

FAN SYSTEMS

When air moves in a ventilation system it does so against resistance. The resistances act in series, e.g. resistances associated with inlets and outlets.

Each system component has a characteristic specific resistance coefficient, k. Few coefficients have been measured for fan ventilated poultry system components, though Pearson and Owen (1994) have published k values for a number of components of natural ventilation systems.

Resistance varies with the square of air velocity (Randall and Boone, 1994). This is a very important point leading designers to attempt to ensure that when the air is expected to negotiate a serious impediment, or a change of direction, it is allowed to do so at as low a velocity as is convenient. Thus, in poultry systems the size of assemblies such as air inlets is normally a compromise between small openings to optimise light and wind proofing, and large openings for minimum flow resistance and maximum fan performance.

Although measurements of k for poultry systems have been rarely made, some simple rules may be deduced from the effect of velocity on resistance. Resistances are likely to be lower for gentle curves than for sharp corners. They are also likely to be lower for smooth surfaces than for rough or obstructed surfaces. Wide openings should generally give better air flow than narrow openings, even at the same air speed, due to edge effects. The air speed at an opening is given by:

$$v = V/A$$

where v = velocity, m/s
V = volume flow, m^3/s
A = cross sectional area, m^2

Resistances are measured in terms of the resulting pressure drop at a particular ventilation rate, in Pa (Newtons/m^2) (based on Randall and Boone, 1994):

$$P = 0.5\rho k v^2$$

where P = resistance, Pa
ρ = density of air, kg/m^3, (Kaye and Laby, 1959, quoted values for ρ of, for example, 1.247 at 10°C and 1.205 at 20°, both for dry air at 760 mm Hg)
v = air velocity, m/s
k = specific resistance coefficient

Fan manufacturers normally supply data for the volume of air flow provided by their fan against a specified pressure in Pa. Many typical modern British poultry ventilation systems operate at about 50 to 75 Pa. It is very important to calculate the numbers of fans required to provide a particular ventilation rate at the appropriate resistance and not at free air (P = 0) because the differences are significant in terms of system performance. Randall and Boone (1994) provided a more detailed account of the relevant theory.

The application of these principles has led to a set of guidelines which have been used in the poultry industry for some years. Many are based on the publications and advice of the early pioneers such as Payne (1959), Tilley (1964) and Prosser (1969). An engineering text commonly used during the early years of the development of controlled environment systems for poultry was Osborne and Turner (1960). Examples were also published by Charles *et al.* (1994).

FREE CONVECTION DRIVEN VENTILATION SYSTEMS

The density of warm air is less than that of cool air. Therefore upward currents occur over a heat source such as a population of birds in a house. In both natural ventilation systems and in failsafe systems for fan ventilated houses it is this heat input which is the driving force for natural ventilation at low wind speeds. The convectional airflow is a function of the square root of the difference in height between inlet and outlet, though strictly it is not quite that simple. Many designs in the past 20 years or so have been based on versions of a formula published by Bruce (1978). In practice it is difficult to design light proofed inlets and outlets for natural ventilation systems, because the openings are necessarily large. However some natural systems with fully automated inlet control appear to have provided temperature control as good as that achieved in powered systems.

WIND EFFECT

Wind is a useful, though variable and uncontrollable, driving force in natural ventilation. It causes much larger pressure differences across buildings than is generally realised, and therefore its potential to interfere with intended ventilation rates in powered ventilation systems is considerable and not always fully appreciated (Randall and Boone, 1994). Consequently, it can be a cause of imprecision in house temperature control, with implications for feed conversion. A great deal of effort has been expended on attempting to build as much wind proofing as possible into system designs.

AIR DISTRIBUTION

Ventilation systems are often judged on the basis of their ability to provide uniform air distribution within a building. A great deal of trial and error has taken place in attempts to achieve uniformity, but there are some physical principles.

Carpenter (1981) described the behaviour of air entering an orifice. He explained that a jet of air widens and slows as distance from the entry point increases, and as house internal air becomes entrained and mixes with it. An isothermal jet (i.e. a jet at the same temperature as the internal air) travels horizontally, but a cool jet (when outside temperature is below inside temperature) gradually falls, and the rate of fall depends on the density difference. The control and adjustment of air inlets depends upon the consideration of these effects. For eaves inlets operating at low air speeds (<4 m/s) in cold weather it is important to aim the incoming air upwards otherwise it will fall rapidly. Too low an entry speed causes inadequate air supply at a distance from an upward tilted hopper type eaves inlet and too high an entry speed causes inadequate supply near it.

Randall (1981) pointed out that when the speed of the air is comparatively fast then the stability of the airflow pattern is characterised by the Archimedes number (the ratio of the buoyancy to the dynamic pressure). A horizontal jet does not fall if the Archimedes number is below 30, but it falls if the number is greater than 75. In practice an Archimedes number below 30 can be achieved by using an inlet speed of at least 5 m/s. The high speed jet ventilation system (HSJ) described by Randall (1981), and further developed for practical application at ADAS Gleadthorpe, uses a jet at 5 to 7 m/s aimed along a smooth ceiling surface. The jet was shown to adhere to the surface in a stable pattern despite changing temperature differences. The small inlet size of this system contributes to wind proofing and light proofing. The original versions generally had eaves inlets, protected from wind, with the two incoming air streams meeting and mixing in the middle of the house, above bird level. Various types of ceiling inlets later also became popular. The attachment of the jet to a smooth surface is called the coanda effect, and it has a considerable influence on the flow pattern in the rest of the building.

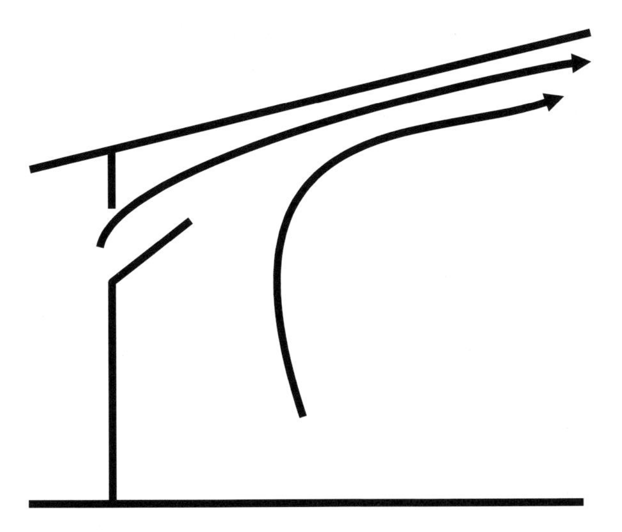

Coanda effect and entrainment

Testing the coanda effect in an empty building with a simple smoke generator

An example of a simple motorised inlet air delivery hopper in the early 1980s at ADAS Gleadthorpe, Nottinghamshire, UK. Note the failsafe panel for the provision of air supply by buoyancy forces in emergencies.

Carpenter (1981) described the typical patterns of airflow likely to occur in pig and poultry buildings as a result of combinations of the above forces with the heat of the animals and the obstructions such as cages and pig pens. Rotary patterns tend to occur, with air rising over heat sources and entrained by primary inlet jets. Randall and Boone (1994) listed the general rules for the prediction of the form of air patterns, quoting Randall (1975).

Air within a building moves in rotary patterns. A primary flow path runs from the inlet to the outlet and secondary paths are induced by it. Obstacles in the path of an airstream can change its pattern and direction. At high ventilation rates the pattern is initially established by the inlet, but modified by obstructions. At low ventilation rates thermal buoyancy causes slow incoming jets to fall.

Some types of ventilation system

FREE CONVECTION SYSTEMS

These are not popular in the UK intensive poultry industry because of the difficulty of light proofing and wind proofing. In contrast, fully automated versions called ACNV (automatically controlled natural ventilation) are common in the pig sector where light proofing is less important. An advantage is the lack of fan noise.

FAN EXTRACTION SYSTEMS

There are many configurations of these, but generally those employing high inlet speeds have been the most successful. The high speed jet system has versions with the inlets at the eaves or with the inlets in a horizontal ceiling. Occasionally the inlets are mounted at the ridge. Most modern versions have automated inlets, linked to the fan controls, so that the design inlet speed is maintained despite variations in ventilation rate.

It has been shown at ADAS Gleadthorpe and elsewhere (e.g. Randall, 1981) that in high speed jet systems the air distribution is inlet dominated. Thus, there is a wide choice of fan positions. The high speed jet system has generally been associated with good temperature control and air distribution pattern, but there can be a poorly ventilated zone adjacent to the side walls under the air inlets of eaves inlet versions.

PRESSURISED INPUT SYSTEMS

There are also many systems in which the fans blow inwards instead of extracting. Such systems may be slightly less noisy than extraction systems since the fan noise can be partially absorbed within the building. Some versions have given good control and air distribution, but not all. Duct design often requires specialised knowledge, though there are published designs available (e.g. Randall, 1981).

PACKAGE SYSTEMS

There are now several packaged and modular ventilation systems on the UK and European markets, most of which are based on automatic inlet control, and some have moulded fan box assemblies and inlet modules for smooth air flow.

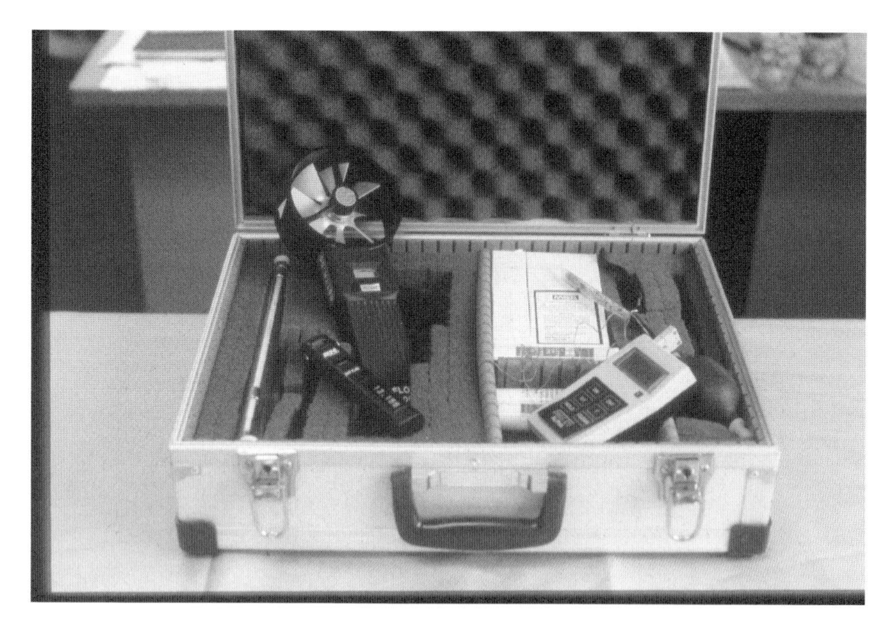

A poultry consultant's testing kit for house environment

Note the precision fit and draught excluding strip, in order to aim the incoming jet in its intended trajectory only

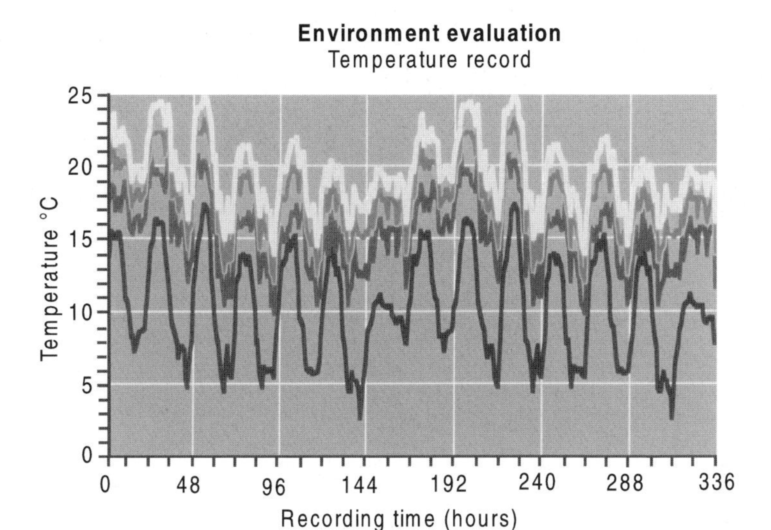

Monitoring inside temperature against outside in a poorly controlled house

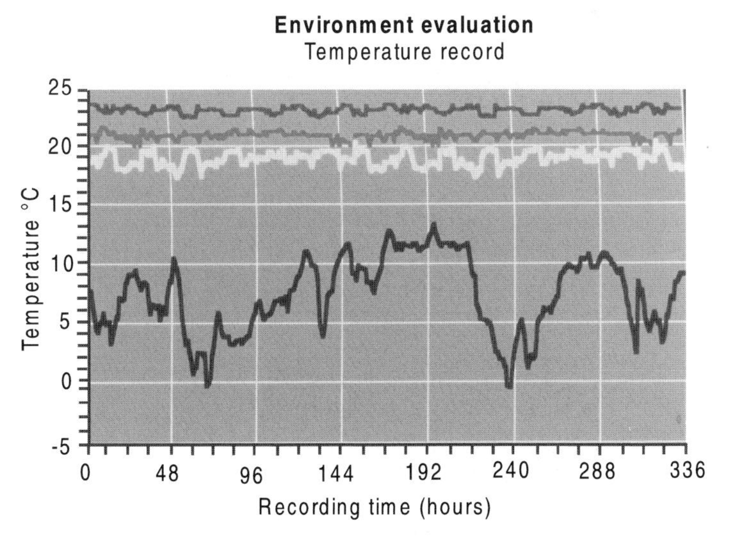

Inside temperature against outside temperature in a well controlled house

A failsafe drop out panel at ADAS Gleadthorpe

An example was described by Findhammer (1994) in which the control of ventilation rate was shown to be accurate.

Environmental monitoring

TEMPERATURE

Adequate measurement of the thermal performance of a poultry house may require data logging of inside temperature against outside shade temperature at least once per hour for an extended period, e.g.14 days (Sutcliffe *et al.,* 1987). The slope and the standard deviation of a linear regression of inside temperature upon outside temperature were shown to provide indices of controllability for houses running without heating or brooding. Good control was defined as an effectively constant inside temperature, having a low standard deviation at outside temperatures over the range approximately -3°C to +18°C, for buildings containing adult layers or broilers after the brooding stage (i.e. buildings with an adequate metabolic heat source). In practice a low standard deviation often indicates good wind proofing.

AIRFLOW PATTERNS

Simple tests using smoke tubes can sometimes provide useful diagnostic information on the patterns of air movement.

AIR QUALITY

The most frequently measured indicator of air quality is the ammonia concentration, which may be tested with colourimetric tubes. Since the ammonia concentration is often the first limiting factor in the setting and operation of the minimum ventilation rate, and since this may therefore make it the limiting factor for temperature control in winter, it is important to measure and control it. Carbon dioxide concentration is sometimes also measured in order to confirm the adequacy of minimum ventilation rates.

AIR CHANGE RATE

It is difficult and expensive to measure the house air change rate precisely, though tracer gas recovery techniques are available. Ventilation rate may be estimated by measuring the air speed across the inlets or outlets, but such measurements are not strictly reliable due to turbulence at the point of measurement. The difference in pressure (Pa) across ventilation apertures may also be used to monitor ventilation rate.

Alarms and failsafes

It is important to provide backup against failure of the ventilation, not only for animal welfare reasons but also for commercial reasons. Failures can include both interruptions to the electricity supply and system failures. In modern large tightly controlled ventilation systems the temperature, humidity, carbon dioxide and ammonia levels would rise quickly if an alternative air supply were not provided. Therefore effort has been put into the specification of standby systems for many years. In the UK the Codes of Recommendations for the Welfare of Livestock (MAFF 1990), the Welfare of Battery Hens Regulations (1987), and the Welfare of Livestock Regulations (1994) have been key influences. These Regulations demand that where automatic equipment includes a ventilation system there must be an alarm (tested not less than once every 7 days) to warn of failure of the system, and there must be additional equipment to provide adequate ventilation in the event of such a failure. Where artificial ventilation is used The Welfare of Farmed Animals (England) Regulations 2000 also call for back up systems and alarms, inspected and tested at least once every 7 days.

Failsafe ventilation systems are driven by free convection. The calculation of the necessary inlet and outlet sizes is therefore based on the buoyancy principles described above.

Targets and summary of recommendations

- The maximum ventilation rate requirement is for removing surplus heat. The minimum ventilation rate requirement is for the provision of air quality. For air quantities see Tables 2.4 and 2.5 above.
- Ventilation systems: A large number of configurations and designs exist. Suitable systems have key characteristics including the following:
    - Light proof if appropriate (see Chapter 4).
    - Draught free (normally less than 0.15 m/s) and uniform air distribution at stock level.
    - Surfaces should be condensation free. For controlled environment housing this normally means insulated walls and roof (to a standard giving a U value of 0.35 W/m^2K or better), and the insulation provided with a vapour check to prevent moisture penetration causing deterioration.
    - Air supply to be controllable between the maximum and the minimum ventilation rates, and stable despite changing outside wind effects.
    - Air quality not to be compromised. This includes keeping CO_2 concentrations below 0.3% and ammonia concentrations as close to zero as possible.
    - Temperature to be controllable within narrow tolerances for laying hens, broiler chickens and controlled environment turkeys (see Chapter 1).
    - The air distribution to be stable against changing outside conditions.
    - An adequate system of emergency ventilation to be included, in order to cope with breakdowns or with the failure of the power supply. Alarms to warn of failure are essential, and mandatory in UK.

References

Anderson, D.P., Beard, G.W. and Hanson, R.P. (1964) The adverse effects of ammonia on chickens including resistance to infection with Newcastle disease. *Avian Disease* **8:** 369-379

Beck. M.M. and DeShazer, J.A. (1984) Biophysical factors affecting energy requirements for poultry production. Project NE127

Becker, A., Vanhooser, S.L. and Teeter, R.G. (1995a) The effect of atmospheric oxygen level and bronchodilator on ascites incidence and performance using day old broiler chickens. 16th Annual meeting Southern Poultry Science Association. *Poultry Science* **74:** 177

Becker, A., Vanhooser, S.L. and Teeter, R.G. (1995b) Effect of oxygen level on ascites incidence and performance of broiler chicks. 16th Annual meeting Southern Poultry Science Association. *Poultry Science* **74:**177

Brody, S. (1945) *Bioenergetics and growth.* Reinhold, New York

Bruce, J.M. (1978) Natural convection through openings and its application to cattle building ventilation. *Journal of Agricultural Engineering Research* **23:** 151-161

Bullis,K.L., Snoeyenbos, G.H. and Van Roekel, H. (1950) Keratoconjunctivitis in chickens. *Poultry Science* **29:** 386-389

Carpenter, G.A. (1981) Ventilation systems. In: *Environmental aspects of housing for animal production* Edit. Clark, J.A., Butterworths, London, 331-350

Carpenter, G.A. (1986) Dust in livestock buildings - a review of some aspects. *Journal Agricultural Engineering Research* **33:** 227-241

Charles, D.R. (1970) Poultry environment in the UK. A review of progress. *World's Poultry Science Journal* **26:** 422- 434

Charles, D.R. (1981) Practical ventilation and temperature control for poultry. In: *Environmental aspects of housing for animal production.* Edit. Clark, J.A. Butterworths, London, 183-195

Charles, D.R. (1993) Air quality problems in poultry production: ammonia, temperature and performance. In: *Biotechnology: recent advances in reducing environmental pollution while improving performance.* Edit. Lyons, T.P., Alltech 7th European Lecture Tour, Alltech Inc., Nicholasville, 11-20

Charles, D.R. (1994) Comparative climatic requirements. In: *Livestock housing*, Edit. Wathes, C.M. and Charles, D.R., CAB International, Wallingford, 3-24

Charles, D.R., Elson, H.A. and Haywood, M.P.S. (1994) Poultry housing. In: *Livestock housing*. Edit. Wathes, C.M. and Charles, D.R., CAB International, Wallingford, 249-272

Charles, D.R and Payne, C.G. (1964) The effects of ammonia on the performance of laying hens. *World's Poultry Science Association 2nd European Poultry Conference*, Bologna

Charles, D.R., Scragg, R.H. and Binstead, J.A. (1981) The effects of temperature on broilers: ventilation rates for the application of temperature control. *British Poultry Science* **22:** 493-498

Charles, D.R. and Spencer, P.G. (1976) *The climatic environment of poultry houses*. Bulletin 212, Ministry of Agriculture, Fisheries and Food, HMSO, London

Clark, J.A. and McArthur, A.J. (1994) Thermal exchanges. In: *Livestock housing*. Edit. Wathes, C.M. and Charles, D.R., CAB International, Wallingford, 97-122

Cropsey, M. (1951) Simplifying poultry ventilation with mathematics. *Agricultural Engineering,* St. Joseph, Michigan **32:** 675

Dann, A.B. (1926) Wet litter in poultry houses. *Poultry Science* **3:** 15-19

Davies, C.N. (1951) Ventilation and its application to poultry housing. *World's Poultry Science Journal* **7:** 195-200

Deaton, J.W., Reece, F.N. and Bouchillon, C.W. (1969) Heat and moisture production of broilers. 2. Winter conditions. *Poultry Science* **48:** 1529-1582

De Shazer, J.A., Olson, L.L. and Mather, F.B. (1974) Heat losses of large white turkeys- 6 to 36 days of age. *Poultry Science* **53:** 2047-2054

Dobson,C., Charles, D.R., Emmans, G.C. and Rhys, I.W. (1969) *Poultry housing and environment*. Bulletin 56, Ministry of Agriculture, Fisheries and Food, HMSO, London

Emmans, G.C. (1974) The effects of temperature on the performance of laying hens. In: *Energy requirements of poultry*. Edit. Morris, T.R. and Freeman, B.M. British Poultry Science Ltd., Edinburgh, 79-90

Emmans, G.C. and Charles, D.R. (1977) Climatic environment and poultry feeding in practice. In: *Nutrition and the climatic environment*. Edit. Haresign, W., Swan, H. and Lewis, D., Butterworths, London, 31-50

Emmel, M.W. (1941) A poultry ventilation problem. *Agricultural Engineering* **22:** 435- 436

Esmay, M.L. (1958) Temperature and humidity control inpoultry houses. *Poultry Science* **37:** 1201-

Findhammer, J.A.A. (1994) Environmental control related to pollution and economics. *World's Poultry Science Association, Proceedings 9th European Poultry Conference,* Glasgow, 272-275

Gavin, A., Konarzewski, M., Wallis, I. and McDevitt, R. (1998) The relationship between metabolic rate and organ size in two strains of chicken. *World's Poultry Science Association (UK Branch) Proceedings of spring meeting*, Scarborough, 105-106

Gordon, W.A.M. (1963) Environmental studies in pig housing. IV. The bacterial content of air in piggeries and its influence on disease incidence. *British Veterinary Journal* **119:** 263-273

Grub, W., Rollo, C.A.,and Howes, J.R. (1965) Dust problems in poultry environments.*Transactions American Society of Agricultural Engineers* **8:** 338-339

Harry, E.G. (1964) The survival of *Eschericia coli* in the dust of poultry houses. *Veterinary Record* **76:** 466-470

Hawk, W. (1910) *Poultry keeping for profit.* Cornwall County Council, Helston, 23

Haywood, M.P.S. (1990) Unpublished ADAS data

Health and Safety Executive (1992) *Occupational exposure limits.* Environmental Hygiene 40/92, HMSO, London

Hill, S.R. (1951) A consideration of the problems involved in ventilating the poultry laying house. *Poultry Science* **30:** 558-568

Kaye, G.W.C and Laby, T.H. (1959) *Tables of physical and chemical constants*, Longmans, Green and Co., London, 132

Kleiber, M.(1961) *The fire of life*. Wiley, New York

Koon, J., Howes, J.R., Grub, W. and Rollo, C.A. (1963) Poultry dust:origin and composition. *Agricultural Engineering* 608-609

Larbier, M. and Leclercq, B. (1994) *Nutrition and feeding of poultry.* Translated and edited Wiseman, J., Nottingham University Press and INRA

Longhouse, A.D., Ota, H. and Ashby, W. (1960) Heat and moisture data for poultry housing. *Agricultural Engineering* St. Joseph, Michigan, **41:** 567

Longhouse, A.D., Ota, H., Emerson, R.E. and Heishman, J.O. (1968) Heat and moisture design data for broiler houses. *Transactions of the American Society of Agricultural Engineers* **11:** 694-700

Lundy, H., MacLeod, M.G. and Jewett, T.R. (1978) An automated multi-calorimeter system: preliminary experiments on laying hens. *British Poultry Science* **19**: 173-186

MacLeod, M. (1990) Energy and nitrogen intake, expenditure and retention at

20° in growing fowl given diets with a wide range of energy and protein contents. *British Journal of Nutrition* **64:** 625-637

MacLeod, M.G., Tullett, S.G. and Jewett, T.R. (1980) Circadian variation in the metabolic rate of growing chickens and laying hens of a broiler strain. *British Poultry Science* **21:** 155-159

Ministry of Agriculture, Fisheries and Food (1990) *Codes of recommendations for the welfare of livestock.* MAFF Publications, London

Ministry of Agriculture, Fisheries and Food (1987) *The welfare of battery hens.* Regulations. MAFF Publications, London

Ministry of Agriculture, Fisheries and Food (2000) *The welfare of farmed animals (England).* Regulations. MAFF Publications, London

Ministry of Agriculture, Fisheries and Food (1994) *The welfare of livestock regulations* 1994. Statutory Instrument No.2126

Mitchell, M.A. (1996) Ascites syndrome: a physiological and biochemical perspective. *Proceedings World's Poultry Science Association (UK Branch) Spring Meeting,* Scarborough

Mitchell,H.H. and Kelley, M.A.R. (1933) Estimated data on the energy, gaseous, and water metabolism of poultry for use in planning the ventilation of poultry houses. *Journal of Agricultural Research* **47:** 735-748

Monteith, J.L. (1984) Consistency and convenience in the choice of units for agricultural science. *Experimental Agriculture* **20:** 105-117

Olson, L.L., De Shazer, J.A. and Mather, F.B. (1974) Convective, radiative and evaporative heat losses of white leghorn layers as affected by bird density per cage. *Transactions American Society of Agricultural Engineers* **17:** 960-967

Osborne, W.C. and Turner, C.G. (1960) *Woods practical guide to fan ventilation engineering.* Woods of Colchester Ltd.

Oyetunde, P.P.F., Thomson, R.G. and Carlson, H.C. (1978) Aerosol exposure of ammonia, dust and *E.coli* in broiler chickens. *Canadian Veterinary Journal* **19:** 187-193

Payne, C.G. (1959) Air conditions in the broiler house. *Agriculture*, London, 629-633

Payne, C.G. (1961) Studies on the climate of broiler houses. 1. Air movement. *British Veterinary Journal* **117:** 36-43

Payne, C.G. (1967) Factors influencing environmental temperature and humidity in intensive broiler houses during the post brooding period. *British Poultry Science* **48:** 1297-1303

Pearson, C.C. and Owen, J.E. (1994) The resistance to air flow of farm building ventilation components. *Journal of Agricultural Engineering Research* **57:** 1-13

Prosser, H.W. (1969) Poultry house ventilation. *Proceedings World's Poultry Science Association (Australian Branch) meeting*, Queensland

Pugh, M. (1978) *The energy balance of growing ducks*. BSc thesis, University of Nottingham

Randall, J.M. (1975) The prediction of airflow patterns in livestock buildings. *Journal of Agricultural Engineering Research* **20:** 199-215

Randall, J.M. (1981) Ventilation system design. In: *Environmental aspects of housing for animal production*. Edit. Clark, J.A., Butterworths, London, 351-369

Randall, J.M. and Boone, C.R. (1994) Ventilation control and systems. In: *Livestock housing*. Edit. Wathes, C.M. and Charles, D.R., CAB International, Wallingford, 149-182

Reece, F.N., Deaton, J.W. and Bouchillon, C.W. (1969) Heat and moisture of broilers. 1. Summer conditions. *Poultry Science* **48:** 1297-1303

Richards, S.A. (1974) Aspects of physical thermoregulation in the fowl. In: *Heat loss in animals and man*. Edit. Monteith, J.L. and Mount, L.E., Butterworths, London, 255-276

Richards, S.A. (1977) The influence of loss of plumage on temperature regulation in laying hens. *Journal of Agricultural Science*, Cambridge **87:** 393-398

Robinson, L. (1948) *Modern poultry husbandry*. Crosby Lockwood, London, 375

Sainsbury, D.W.B. (1959) *Electricity and problems of animal environmental control*. Electrical Development Association. Publication No. 1897

Saville, C.A., Clark, J.A. and Charles, D.R. (1978) House heat balance. Paper to *World's Poultry Science Association (UK Branch)*, London

Scorgi, N.J. and Willis, G.A. (1952) In: *Linton's Veterinary Hygiene*. Green, Edinburgh

Sutcliffe, N.A., King, A.W.M. and Charles, D.R. (1987) Monitoring poultry house environment. In: *Computer applications in agricultural environments*. Edit. Clark, J.A., Gregson, K. and Saffell, R.A., Butterworths, London, 207-218

Tilley, M.F. (1964) Climate in livestock buildings. *Proceedings E.D.A. Rural electrification conference*, University of Nottingham

Trotter, W. (1851) Essay on the rearing and management of poultry. *Journal of the Royal Agricultural Society of England* **12:** 161-202

Tucker, S.A. and Walker, A.W. (1992) Hock burn in broilers. In: *Recent advances in animal nutrition*. Edit. Garnsworthy, P.C., Haresign, W. and Cole, D.J.A., Butterworth Heineman, Oxford, 33-50

Valentine, H. (1964) A study of the effect of different ventilation rates on the ammonia concentrations in the atmosphere of broiler houses. *British Poultry Science* **5:** 149-160

von Wachenfelt, E., Pedersen, S. and Gustafsson, G. (2001) Release of heat, moisture and carbon dioxide in an aviary system for laying hens. *British Poultry Science* **42:** 171-179

Wathes, C.M. (1978) *Sensible heat transfer from the fowl*. PhD thesis. University of Nottingham

Wathes, C.M. (1981) Insulation of animal houses. In: *Environmental aspects of housing for animal production*. Edit. Clark, J.A., Butterworths, London, 379-412

Wathes, C.M. (1994) Air and surface hygiene. In: *Livestock housing*. Edit. Wathes, C.M. and Charles, D.R., CAB International, Wallingford, 123-148

Wathes, C.M. (1998) Aerial emissions from poultry production. *WPSA (UK Branch) Proceedings of Spring Meeting,* Scarborough

Webster, A.J.F (1981) Optimal housing criteria for ruminants. In: *Environmental aspects of housing for animal production.* Edit. Clark, J.A., Butterworths, London, 217-232

Wilson, E.A. (1986) *Surface energy exchange in poultry houses*. BSc thesis, University of Nottingham

Xin, H., Sell, J.L. and Ahn, D.U. (1996) Effects of light and darkness on heat and moisture production of broilers. *Transactions American Society of Agricultural Engineers* **39:** 2255-2258

3

PREVENTING AND ALLEVIATING HEAT STRESS IN BROILERS

D.R.Charles and the late S.A.Tucker

The problems

There is much more to coping with heat stress than preventing mortality. Heat stress is an animal welfare issue. Growth rate and profitability are affected at temperatures well below those which cause mortality. The optimum temperature for performance in the finishing phase, depending on feed cost and broiler price, is likely to be in the range 18°C to 22°C, (Charles *et al.*, 1981; Wathes *et al.*, 1981; Charles, 1986; Charles, 1994) (See Chapter 1).

Physiological background

Mount (1974) described a band of temperature, called the thermoneutral zone, within which animals need to expend least energy to either keep warm or to keep cool. In order to attempt to maintain a more or less constant body temperature of approximately 41°C chickens pant at high environmental temperatures after a rise in hypothalamic temperature (Richards, 1974). Teeter and Belay (1996) estimated that the respiration rate may increase by 10 fold. Once panting sets in evaporative heat loss becomes dominant over sensible (convective and conducted) heat losses (Richards, 1974), and this is why it is so much more difficult to provide bird comfort in hot humid climates than in hot dry climates.

The increase in respiration rate during panting leads to a loss of carbon dioxide and a rise in plasma pH, (Etches *et al.*, 1995). This condition is known as respiratory alkalosis. The needs of heat loss conflict with those of alkalosis, and birds have systems to minimise this conflict by using airways not involved in respiration, one of which is called gular flutter. High temperatures are eventually associated with increased plasma sodium and chloride, and decreased potassium and phosphate. The need for water increases during heat stress.

Acclimatisation, or the acquired ability to withstand slightly higher temperatures, takes 3 to 5 days in the adult chicken. Blood flow to the combs, wattles and shanks increases during heat exposure, which means that broilers may be more susceptible than layers. Emmans and Dun (1973) found at ADAS Gleadthorpe that layers exposed to 24°C for 12 months had more comb and wattle weight per kg liveweight than those held at 19°C, and that the effect was more pronounced in a white breed than in a brown. Acclimatisation is an important survival mechanism. It would therefore be useful, though perhaps debatably practicable, to anticipate very severe hot weather by running the finishing temperature towards the high end of the optimal range for a few hours per day for a few days before a heat wave. Experience in temperate climates indicates that the first hot weather of the year, particularly after a cold spring, is often the most challenging, but this may merely be because growers are less prepared than later in the season. Teeter and Belay (1996) stressed the importance of acclimatisation. Zhou *et al.* (1997) found that prior exposure to high temperatures at 5 days was helpful in coping with the challenge of 33°C for 3 hours at market weight.

While blood flow is diverted to the perifery during heat stress digestion is less well served. Non-evaporative cooling is increased by wing spreading, (Teeter and Belay, 1996).

Several important hormonal mechanisms are involved in thermoregulation, and have been summarised by Etches *et al.* (1995). Melatonin may alter the set point of the hypothalamus, which is effectively the site of the body thermostat. It is interesting to speculate whether this melatonin involvement accounts for any of the effects of lighting patterns on the response of broilers to hot weather. Heat stress reduces the levels of luteinising hormone (LH) in adult females, and of the preovulatory surges of LH and progesterone. Levels of thyroid hormone decrease.

A response of all organisms, including birds, to high temperature is the increased synthesis of heat shock proteins, which to some extent protect heat sensitive proteins, (Etches *et al.*, 1995). It might be useful to discover the essential amino acid composition of chicken heat shock proteins, in case there is any need to allow for their synthesis in summer feed formulation. Gabriel *et al.* (1996) found that heat shock protein 70 messenger RNA and heat shock protein 70 increased in broiler liver during heat stress. They considered that heat shock protein may be involved in the acclimatisation process.

Acute heat stress can cause muscle damage by an increase in the plasma activity of muscle creatine kinase isoenzyme (CK), (Mitchell and Sandercock, 1995). Yahav *et al.* (1994) found that plasma triiodothyronine levels were related to thermal challenge.

Despite the undoubted importance of these physiological mechanisms, and of some of the biochemical effects reviewed below, (see *Practical alleviation of the problem*, section *8. Nutrition,* below), it is probably important to distinguish between true stress, indicated by panting, and thermal depressions in performance, mainly attributable to the depression in feed intake. Thus Al-Harthi and MacLeod (1996), in a paired feeding test in calorimeters, were able to attribute the difference in weight gain between birds at 20°C and 30°C entirely to feed intake, by feeding the birds at 20° the same amount that those at 30°C had eaten on the previous day. But for stressed birds Dale and Fuller (1980) found only 63% of the effect on growth rate to be attributable to feed intake.

Genetic aspects

There are genetic differences in heat tolerance. White Leghorns are often considered to be more tolerant than the heavier breeds, (Gowe and Fairfull, 1995), and some tropical breeds such as the Bedouin are more tolerant still. In addition to the difference between breeds in the weights of combs and wattles reported by Emmans and Dun (1973) it was also found at ADAS Gleadthorpe that a white layer breed of the time had less weight of feathers per unit of body weight than a contemporary brown breed.

There are some single genes relevant to heat tolerance, such as those associated with the naked neck and frizzled plumage conditions, (Gowe and Fairfull, 1995). Selection for leanness and feed efficiency may have increased heat tolerance. Yalcin *et al.* (1997a) noted significant season x stock interactions for weight gain when testing three stocks in Turkish hot and temperate seasons. Yalcin *et al.* (1997b) found that naked neck (Na/na) stocks had an advantage in hot conditions.

Physical aspects

For both buildings and animals Clark and McArthur (1994) have provided a detailed account of the physics of heat balance. Heat transfer to and from a building takes place in the ventilation air, through the structure of the walls and roof, and through the floor. There is also heat storage in the structure and by the birds. Heat balance equations have been developed by a number of authors. A simple equation for practical poultry house applications was published by Saville *et al.*, (1978). It generally gives good predictions for houses at equilibrium. Its use in practical work, in the following format, was described by Charles (1981)(see also Chapter 2).

$$\Delta T = Q_s/(1200V + UA)$$

where: ΔT = temperature lift above outside temperature, °K

Q_s = sensible heat output, W/bird (e.g. about 8.8 for a well feathered 2.3 kg layer at 20°C, (Richards, 1977), and 5.5 for a 1.73 kg broiler in large groups at 17°C, (Wathes, 1978), or 9.6 for a 2.04 kg broiler at 19°C, (Longhouse *et al.*, 1968)). See below for further definitions of sensible heat (see also Table 2.2).

V = ventilation rate, m^3/s per bird

U = average thermal transmittance of the walls and roof, W/m^2K

A = exposed area of walls and roof, m^2/bird, (normally greater than the floor area by a factor of 1.2 to 1.7, depending on the shape of the building)

1200 = Approximate volumetric heat capacity of air at usual air temperatures and humidities, J/m^3K

Note that this equation does not include a term for heat storage, and therefore it is only predictive at equilibrium, (see also Chapter 2 for other sources of imprecision). Buildings may not be at equilibrium during a few hours of rapid heating up during very hot days. Sampling and data logging procedures for monitoring poultry house environments, and relating the recordings to this equation, were described by Sutcliffe *et al.,* (1987) and are discussed in Chapter 2.

Birds must also maintain a balance between heat production and heat loss. The principles of the heat balance of animals are well known and have been reviewed in depth by Monteith (1974) and by Blaxter (1989). Heat production comes from metabolism. Heat is lost by convection, evaporation and radiation, and a little by conduction. In medium and long term the total of these components must equal the rate of heat production, though a small amount of storage takes place. The conduction, convection and radiation components are conventionally grouped together and called the sensible heat loss, because it can be sensed, and it is this heat which causes the internal building temperature to rise above outside. The evaporative losses are very important to the bird in hot weather, but they do not heat the building. These factors were reviewed by Charles (1992).

Practical alleviation of the problem

VENTILATION ADEQUACY

The house heat balance equation indicates that the temperature inside a poultry building without cooling equipment will normally be above the outside temperature. Exceptions are periods of warming up or cooling down when equilibrium has not yet been reached. The greater the mass of the building, its immediate surroundings (i.e. concrete gangways etc.), and its contents, the longer the equilibration takes, thus delaying the onset of peak temperatures. This is why in hot countries stone or concrete structures are sometimes used, and it is why it may be possible to achieve some useful overnight cooling of the site, lasting well into the day.

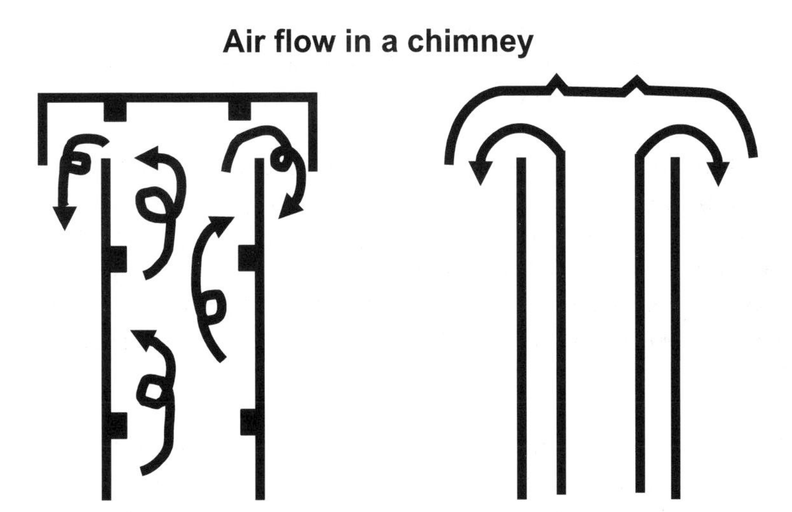

Some sources of fan performance depression (left) compared with smooth air flow (right)

By arbitrary convention British designers have defined the maximum capacity of the ventilation system as the air change rate necessary to prevent the temperature rise above outside shade temperature exceeding approximately 3°C (see also Chapter 2). This recommendation is a conversion from the old 5° to 6°F, since it goes back to the early development of controlled environment systems. Charles (1970) reviewed 6 references on the subject from 1951 to 1961, in particular Sainsbury (1959), who recommended a maximum ventilation rate of 1.5 to 2 cubic feet per minute per pound liveweight (cfm/lb), (1 m^3/s approximately = 2118 cfm). Current recommendations are given in Chapter 2. It is very important to stress that all calculations of the air provided, when choosing fans, should be net after allowing for system resistance, which is normally at least 50 Pa in powered ventilation systems (see Chapter 2).

Solutions of the equation show, however, that the temperature lift, ΔT, is relatively insensitive to small changes in V at high values of V. This contrasts starkly with its very high sensitivity at low values of V. In practice there are two issues. The first is the need to install sufficient air change capacity, and the second is the need to avoid installation faults, such as unnecessary obstructions to airflow. MAFF (1993) provided some diagrams of typical examples of installation difficulties.

Ventilation recommendations assume that all of the birds get an approximately equal share of the air, and it is the function of the air distribution system to ensure that this happens. In many systems it can be useful to assist the distribution with internal circulation fans. Their air *movement* should not be confused with the air *change* rate provided by the main fans. They are normally cheap to run since they operate at low resistance, though they have a capital cost.

INSULATION

In hot weather the function of insulation, particularly in the roof, is to minimise solar heat penetration. For this purpose MAFF (1993) recommended a standard equivalent to U = 0.4 W/m^2 °C or lower, where a low value indicates good insulation. It is now known that even in the climate of Britain solar penetration through the roof can be substantial. Wilson (1986) found solar gains of 30 W/m^2 coming through old broiler house roofs. Haywood (1990) found that the solar gain through single skin cattle shed roofs could be as much as 85 W/m^2.

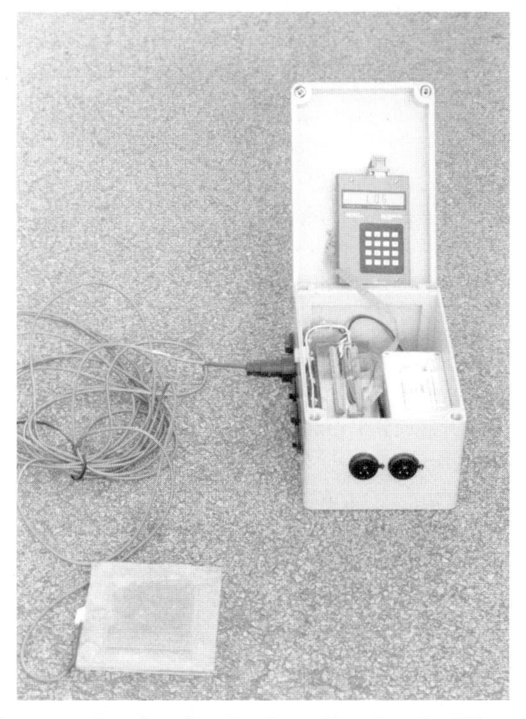

Measurements of solar penetration of roofs and walls, such as those by Wilson (1986) and Haywood (1990), used flux plates

Roof sheet colour, reflectivity, pitch and orientation all have effects on the solar gain, though usually small compared with that of insulation. Owen (1994) quoted data showing that 10 year old aluminium is only 97% as effective as the new material when used for shading. Cooling the roof sheets with water can be helpful, if permitted by the local water supplier and if the water is not allowed to raise the relative humidity near the air inlets, (see below).

Smith (1981) pointed out that the absorbed energy is converted into heat and conducted to cooler parts of the structure or emitted as long wave radiation. Thus materials for the outside of the building absorb less if they have a low absorptivity for short wave radiation (usually written α) and and a high emissivity for long wave radiation (ε). Materials with a low ratio between these two factors (α/ε) will therefore be advantageous for minimising solar penetration. Anderson (1977) classified materials according to the ratio α/ε, from <0.5 (Class 1; example white paint) through to $= >1$ (Class 4; example black paint). Most practical building surfaces and colours used in UK fall within Class 3 (α/

$\varepsilon = 0.8$ to 1.0). Data from Oke (1978) showed that the short wave reflectivity, or albedo (the complement of absorptivity, $(1-\alpha)$), for red, brown and green paints were rather similar at 0.20 to 0.35, whereas black was quoted as 0.02 to 0.15 and white or whitewash was 0.50 to 0.90.

CIRCULATION FANS AND AIR SPEED

Internal circulation fans assist with air distribution, and they also provide some direct cooling for the birds by increasing the rate of convectional heat loss. Some ventilation systems use the main air change fans to provide high air speeds over the birds.

Around the animal body there is a thin boundary layer of still air from which convection takes place by laminar flow. Movement of the external air can thin the boundary layer and increase heat loss. Wathes (1978) measured the effect of air speed across the surface of a feathered pelt in a wind tunnel on the physical characteristics of the boundary layer, and on the transition between natural and forced convection. He calculated that the change occurred at 0.1 to 1 m/s, depending on the temperature gradient from the bird surface to the air. Mitchell (1985) estimated the effects of air speed on heat loss and recommended the use of air speeds of 0.3 to 1 m/s over the birds. At 30°C the relationship was described by:

$$y_1 = 11.8 + 40.1\,x$$

where y_1 = convective heat loss, W/m^2
x = air speed, m/s

also $y_2 = 41.4 + 25.9x$

where y_2 = sensible heat loss, W/m^2

Seven week old female broilers at 30°C and 18°C dew point (about 47% relative humidity) when cooled by air moving at 2 m/s, had the same growth rate as birds at 26°C in still air (<0.25 m/s) (Lott *et al.,* 1998).

MAFF (1993) described practical methods of achieving high air speeds over birds. Runge (1991), working under the climatic conditions of Queensland, Australia, found that substantial numbers of cooling fans are needed to achieve high air speeds in practical broiler houses. Bottcher *et al.,* (1995) measured the air speed at 25 cm above floor level as affected by fan type, height of mounting and tilt angle. Each make of fan seemed to have its own characteristics, so the data may not be universally applicable, but the principles are of interest. For tilt angles of 20° or less from the horizontal, area averaged velocity increased with tilt angle and decreased with increasing fan height. Angles of 10° to 14° were often satisfactory.

EVAPORATIVE COOLING

Evaporative cooling is an extremely useful technique in hot dry climates. Bolla (1991) described practical installations in Australia, and the choice of spray nozzles for medium to high pressure foggers. In his work the spray nozzles were normally mounted close to side wall inlets.

Evaporation can only be used with extreme care in UK, since there is a risk that despite a drop in dry bulb air temperature the increase in relative humidity may be harmful. The rate of respiratory and cutaneous evaporative losses from the birds are proportional to the vapour pressure gradient between the tissues and the air. Most of the evaporation takes place from the respiratory tract, hence the importance of panting. Evaporative water loss decreases by 0.7 g/hour per kg liveweight per kPa increase in vapour pressure, (Richards, 1976). Kettlewell and Moran (1990) predicted that the maximum increase in respiratory rate could occur at a temperature as low as 27°C in saturated air. Therefore, Charles (1992) recommended that evaporative cooling should only be used if the house relative humidity (rh) can be kept below 70%.

More recently however evidence has appeared suggesting that the relationship between growth and relative humidity at high temperatures may not be a simple matter of keeping the humidity as low as possible. Yahav *et al.* (1994) found that the growth of broilers from 5 to 8 weeks of age at 35°C, after a one week acclimatisation period, was fastest at 60 to 65% rh out of a set of four ranges tested. However with turkeys of 4 to 8 weeks of age growth was fastest at the lowest range tested (40 to 45%). Yahav (2000) tested a range of relative humidities from 40% to 75% at 28 and 30°C. Male broilers from 4 to 8 weeks of age grew fastest at 60 to 65%. Teeter and Belay (1996) recommended that foggers were of little value if the relative humidity was above 60 to 70%. Maloney (1999), in a review of heat balance, recommended minimising humidity if the ambient temperature is above 30°C, as originally recommended by Romijn and Lokhorst (1966). It must be remembered that humidity may have other effects relevant to growth, such as air hygiene effects (Wathes, 1994), apart from its thermoregulatory effects (see Chapter 2).

The principles of ventilation rate requirement, as discussed in Chapter 2, have been worked out in temperate and cold climates. Gates *et al.* (1991) made the point that although it seems common sense to use the maximum ventilation rate during evaporative cooling, they found evidence that it can sometimes be more effective to use less than the maximum. When the outside temperature is above the required temperature, and if the ventilation rate exceeds the capacity of the evaporating equipment, then hot air is being taken into the building unnecessarily.

STOCKING DENSITY

In view of the importance of the boundary layer (see above) and of conductive and radiative transfer of heat between birds, quite apart from house heat balance effects such as the load on the ventilation system, the common practice of thinning in hot weather is sensible. Wathes and Clark (1981) found that the sensible heat loss of broilers huddled in groups was only 60% of that for isolated birds.

CYCLING HOUSE TEMPERATURE

Cool nights are beneficial during hot weather because they give the building structure and the surroundings time to cool down, but the early work, mainly at Mississippi, (e.g. Reece *et al.,* 1971), reviewed by Charles (1986) and by Sutcliffe *et al.,* (1987), also showed some biological benefits of cycles. The results of experiments on cycles are confusing, however, because it is difficult to distinguish between the effects of cycles *per se* and of the possible stress at the extremes of the cycle. In some experiments the effects of a cycle of moderate amplitude on performance seemed to resemble the effects of the mean temperature, (e.g. Deaton *et al.,* 1972). If this is so then a moderate cycle is likely to be useful in order to lower the daily average. The potential danger is that the lowering of the average could interfere with acclimatisation and therefore with survival on very hot days.

LIGHTING PROGRAMMES

For various reasons there has been interest lately in lighting patterns for broilers, and in the provision of rather more darkness in each 24 hour period than was conventional in UK for many years. Gordon (1994) reviewed work on sleep and rest, and found that 16 hours of light per day may be sufficient for performance. There may be virtue in providing some of the dark period during the afternoon, since metabolic heat production is lower in the dark. MacLeod *et al.,* (1980) found that heat production was 16% lower in the dark at 0.235 kg, 10% at 0.879 kg and 11% at 1.867 kg, when working with a photoperiod of 23L:1D at 20°C.

NUTRITION

Vitamins and minerals

Vitamin C is the most studied vitamin in relation to problems of high temperature. Daghir (1995a) stated that there is some evidence that under conditions of high temperature some birds and mammals cannot synthesise enough ascorbic acid to replace the losses which occur during stress. Daghir (1995b) reviewed several papers on the use of vitamin C supplementation in hot climates to alleviate heat stress, in some cases with reports of improved growth rate. In some of the work reviewed, ascorbic acid was used with the addition of ammonium chloride and tyrosine, perhaps as a catecholamine synthesis blocker to help maintain high ascorbic acid values.

Several of the sources reviewed recommended adding 1 g ascorbic acid per litre of drinking water during heat waves. Okan *et al.* (1996b) found benefits from vitamin C supplementation in Japanese quail. However Teeter and Belay (1996) found no response to vitamin C. Kutlu and Forbes (1994) found that by associating ascorbic acid supplementation with feed colour chicks were able to regulate ascorbic acid intake to temperature. They chose 10 mg/d in a hot environment and 5 mg/d in a cool. Ozcan and Kutlu (1997) found that ascorbic acid significantly improved carcass yield under heat stress. McKee *et al.* (1997) found that supplemental ascorbic acid influenced the body energy stores used during periods of reduced energy intake. Mahmoud and Edens (1998) found that vitamin C modified the heat stress response by increasing the expression of heat shock protein.

Daghir (1995b) reviewed the literature on acid-base imbalance during heat stress, and found several suggestions for its amelioration, including carbonating the drinking water and the addition of ammonium chloride to either the feed or the water. Since potassium losses were found to be high at high temperature it has been claimed that the addition of potassium chloride to the drinking water reduced mortality. Dietary or drinking water supplementation with sodium bicarbonate at high temperature stimulated both feed intake and liveweight gain, probably through water intake and therefore accompanied by wetter droppings (Balnave and Gorman, 1993). They considered that the response was likely to vary with factors influencing acid-base balance. Teeter *et al.* (1996) found that electrolyte supplementation of the drinking water reduced mortality and improved weight gain during heat stress. Teeter and Belay (1996) found that electrolytes such as KCl in the drinking water improved respiratory efficiency (J/breath) by 30%. Balnave and Brake (2001) demonstrated an improvement due to dietary sodium bicarbonate in the liveweight gain of 3 to 6 week old broilers at 31°C.

At high temperature the calcium:phosphorus ratio was found to be particularly important. The plasma sodium:calcium ion ratio was found to increase at high temperature and a high ratio was thought to impair heat tolerance. Mineral intake was capable of influencing the ratio, though not in a clearly defined way. There could be a connection between these observations and the loss of potassium mentioned above.

Daghir (1995b) also pointed out that in heat waves or in the tropics the storage and transport of vitamin premixes are critical because of the risk of denaturation. The inclusion of antioxidants is particularly important and it may be worth adding the fats just before use. Vitamin stability is improved if storage is without choline, and the choline added separately.

There is some evidence, though mostly in layers and breeders, of higher vitamin A requirement during hot weather, due to interference with absorption, (Daghir, 1995a).

Vitamin E is a physiological antioxidant, serving to inactivate free radicals and preserve the integrity of the endothelial cells of the circulatory system. Requirements increase with stress, including heat stress. In layers Bollinger-Lee *et al.* (1998) showed that vitamin E supplementation can at least partly alleviate the effects of heat stress, perhaps by maintaining the supply of egg precursors. Work reviewed by Daghir (1995a) suggested that heat stress may interfere with the conversion of vitamin D_3 to the active form. However he cautioned that there was still no clear evidence of benefits from adding extra vitamins E and D_3 during heat stress.

Energy source

Carbohydrate is used in the synthesis of body fat at an efficiency of 0.80, whereas lipid is used at an efficiency of 0.96, with less heat given off in the process, Blaxter (1989). This has often led to the suggestion that high fat diets should be helpful in hot weather. Up to the time of the literature review of Charles (1986) the results of experiments on the topic were confusing and inconsistent. However the balance of evidence reviewed by Daghir (1995a), including his own experimental work, was that high fat diets were beneficial in hot weather. Presumably the results must have depended upon how much fat the birds were laying down at the time.

Bonnet *et al*., (1997) reported that the nitrogen corrected apparent metabolisable energy (AME_n) of two feeds was reduced at high temperatures. They recommended the use of high quality oil and protein sources for hot weather.

Methionine source

There have been suggestions that the hydroxy analogue of methionine may be better absorbed during heat stress than the synthetic DL form. Knight *et al.,* (1994) found that 2-hydroxy-4-(methyl-thio) butanoic acid, (HMB), gave better weight gains than DL methionine at 30/32°C. The uptake of the D form is apparently more severely limited than that of the L. HMB is absorbed throughout the tract by passive diffusion, whereas the DL form is absorbed in the small intestine by energy dependent processes. Yet Mitchell and Hunter (1996) found no nutritional advantage on the basis of absorption efficiency or jejunal uptake.

Nutrient supply

It is self evident that growth rate must be limited if amino acid intake is depressed below the levels needed to sustain maintenance and lean tissue development. Thus as feed intake declines with increasing temperature it is necessary to formulate for the expected feed intake. However there is some suggestion in the literature that under conditions of high temperature the choice of energy:protein ratio may not be that simple. Thus several sources reviewed by Charles (1986) reported the need to increase both protein and energy levels at high temperatures. Cowan and Michie (1978) found that weight gain depression above about 23°C was not overcome by high protein diets, up to 308 g/kg. Similarly Cheng *et al.* (1997a and b) recommended that at 29°C or above, broilers should not be offered high protein feeds, which they considered increased heat production. Using free choice feeding Sinurat and Balnave (1986) found evidence of the need for high energy concentration at high temperature.

Feed presentation

The data of Sinurat and Balnave (1986) suggested that broilers have some ability to select between energy and protein at high temperatures.

After eating, heat production rises by an amount called the specific dynamic action of food (SDA), a term going back to the classic metabolism studies of the early years of the 20th century, and explained by Sturkie (1976) and Blaxter (1989) . This rise is probably due to factors such as the act of eating, enzymic effects in the gut, absorption, intermediary storage and cellular activity. In mammals the SDA accounts for about 6% of the energy in sucrose, 13% in fats and 30% in proteins according to Gordon (1982). In man the heat increase in metabolic rate peaks about 1 to 2 hours after a meal and persists for up to 15 hours, (Blaxter, 1989). Thus the practice of meal feeding for broilers, (e.g. Walker, 1995), is likely to be associated with lowest heat production at the end of the interval between meals. Teeter and Belay (1996) found that a 3 hour feed withdrawal prior to heat stress improved survival. In recent summers some UK broiler growers have adopted meal feeding during hot weather. To achieve a benefit to the birds it is necessary to anticipate peak house temperature by at least an hour to allow SDA to fall, but it is unlikely to fall to zero. It must not be forgotten, of course, that growth rate will be constrained by the total daily feed intake.

In Japanese quail Okan *et al.,* (1996a) found that wet feeding increased dry matter intake at high temperature.

Water

Hill (1977) found in layers that water intake and feed intake were related, and that cooling the drinking water by 6°C below ambient air temperature increased the intake of both water and feed. Daghir (1995b) stressed the importance of a liberal supply of water for broilers exposed to high temperatures, and suggested attempting to keep it cool by running supply pipes at least 60 cm under the ground and insulating the header tanks. In hot weather wide deep drinkers are advantageous to the birds, because they can immerse the whole face and gain a cooling effect. Teeter and Belay (1996) found that water cooled to 13°C was beneficial to broilers in heat stress. They also suggested that gently walking the birds increased water intake by 8%. May *et al.* (1997) found that water intake was lower with nipple drinkers than with bell type drinkers in hot conditions and that panting birds had difficulty drinking from nipples mounted high off the floor. However drinker design in UK is necessarily dominated by the need to prevent spillage in cool weather, and consequent wet litter, (Tucker and Walker, 1992).

Zhou *et al.* (1998) published some evidence that glucose in the drinking water was beneficial during heat stress. There were effects on growth rate, plasma viscosity, plasma osmolality and plasma protein concentration.

Some recommendations

- The provision of adequate ventilation capacity, and the provision of high air speeds over the birds, are important during heat stress. The ventilation equipment should be checked for air losses due to obstructions or installation errors.
- Solar heat gain through old poorly insulated roofs should not be underestimated, and its only long term amelioration is by improvement of the U value, though roof wetting can be useful if care is taken not to raise the house humidity.
- Stocking density is relevant.
- During warm weather it is useful to cool the building at night and to give the birds the chance to feed at cool times of day.
- The water supply should be genuinely *ad libitum* and as cool as possible.
- There is probably a case for including vitamin C in the diet for hot weather batches.

- There may be a case for high energy finisher diets as well as for the checking of amino acid intake at the expected level of feed intake.

- There is scope for R&D on several nutritional factors, such as levels of vitamins E and D_3, the use of sodium bicarbonate, the raising of potassium levels, and methionine source. Other R&D suggestions include the nutritional requirements for heat shock protein and genetic work on heat tolerance.

References

Al-Harthi, M.A. and MacLeod, M.G. (1996) Analysis of heat stress effects on growth by pair feeding. *World's Poultry Science Association (UK Branch). Proceedings of spring meeting,* Scarborough, 119-120

Anderson, B. (1977) *Solar energy. Fundamentals in building design.* McGraw-Hill, New York

Balnave, D. and Brake, J. (2001) Different responses of broilers at low, high, or cyclic moderate-high temperatures to dietary sodium bicarbonate. *Australian Journal of Agricultural Research* **52:** 609-613

Balnave, D. and Gorman, I. (1993) A role for sodium bicarbonate supplements for growing broilers at high temperatures. *World's Poultry Science Journal* **49:** 236-241

Belay, T. and Teeter, R.G. (1996) Virginiamycin and caloric density effects on live performance, blood serum metabolite concentration, and carcass composition of broilers reared in thermoneutral and cycling ambient temperatures. *Poultry Science* **75:** 1383-1392

Blaxter, K.L. (1989) *Energy metabolism in animals and man.* Cambridge University Press, 86-119

Bollinger-Lee, S., Mitchell, M.A., Utomo, D.B., Williams, P.E.V. and Whitehead, C.C. (1998) Influence of high dietary vitamin E supplementation on egg production and plasma characteristics in hens subjected to heat stress. *British Poultry Science* **39:** 106-112

Bolla, G. (1991) Poultry housing options for broiler growers. In: *National Poultry Housing Seminars.* Chicken Meat Research Council and Egg Industry Research and Development Council, 28-37

Bonnet, S., Geraert, P.A., Lessire, M., Carre, B.and Guillaumin,S. (1997) Effect of high ambient temperature on feed digestibility in broilers. *Poultry Science* **76:** 857-863

Bottcher, R.W., Magura, J.R., Young, J.S. and Baughman, G.R. (1995) Effects of tilt angles on airflow for poultry house mixing fans. *Applied Engineering in Agriculture* **11:** 721-730

Charles, D.R. (1970) Poultry environment in the UK. A review of progress. *World's Poultry Science Journal* **26:** 422-434

Charles, D.R. (1981) Practical ventilation and temperature control for poultry. In: *Environmental aspects of housing for animal production*. Edit. Clark, J.A. Butterworths, London, 183-195

Charles, D.R. (1986) Temperature for broilers. *World's Poultry Science Journal* **43:** 249-258

Charles, D.R. (1992) Theoretical aspects of poultry housing in hot climates. *World's Poultry Science Association, 19th World Poultry Congress,* Amsterdam,117-120

Charles, D.R. (1994) Comparative climatic requirements. In: *Livestock housing*, Edit.Wathes, C.M. and Charles, D.R., CAB International, Wallingford, 3-24

Charles, D.R., Groom, C.M. and Bray, T.S. (1981) The effects of temperature on broilers: interactions between temperature and feeding regime. *British Poultry Science* **22:** 475-482

Cheng, T.K., Hamre, M.L. and Coon, C.N. (1997a) Effect of environmental temperature, dietary protein, and energy levels on broiler performance. *Journal of Applied Poultry Research* **6:** 1-17

Cheng, T.K., Hamre, M.L. and Coon, C.N. (1997b) Responses of broilers to dietary protein levels and amino acid supplementation to low protein diets at various environmental temperatures. *Journal of Applied Poultry Research* **6:** 18-33

Clark, J.A. and McArthur, A.J. (1994) Thermal exchanges. In: *Livestock housing*.Edit. Wathes, C.M. and Charles, D.R., CAB International, Wallingford, 97-122

Cowan, P.J. and Michie, W. (1978) Environmental temperature and broiler performance: the use of diets containing increasing amounts of protein. *British Poultry Science* **19:** 601-605

Daghir, N.J. (1995a) Nutrient requirements of poultry at high temperatures. In: *Poultry production in hot climates*. Edit. Daghir, N.J., CAB International, Wallingford, 101-124

Daghir, N.J. (1995b) Broiler feeding and management in hot climates. In: *Poultry production in hot climates*. Ed. Daghir, N.J., CAB International, Wallingford 185-218

Dale, N.H. and Fuller, H.L. (1980) Effect of diet composition on feed intake and growth of chicks under heat stress. *Poultry Science* **59:** 1434-1441

Deaton, J.W., Reece, F.N., Lott, B.D., Kubena, L.F. and May, J.D. (1972) The efficiency of cooling broilers in summer as measured by growth and feed utilisation. *Poultry Science* **51:** 69-71

Emmans, G.C. and Dun, P. (1973) Temperature and ventilation rate for laying fowls; supplementary report on feathering. Unpublished ADAS data.

Etches, R.J., John, T.M. and Verrinder Gibbins, A.M. (1995) Behavioural, physiological, neuroendocrine and molecular responses to heat stress. In: *Poultry production in hot climates.* Edit. Daghir, N.J., CAB International, Wallingford, 31-65

Gabriel, J.E., Ferro, J.A., Stefani, R.M.P., Ferro, M.I.T., Gomes, S.L. and Macari, M. (1996) Effect of acute heat stress on heat shock protein 70 messenger RNA and on heat shock protein expression in the liver of broilers. *British Poultry Science* **37:** 443-441

Gates, R.S., Timmons, M.B. and Bottcher, R.W. (1991) Numerical optimisation of evaporative misting systems. *Trans. American Society of Agricultural Engineers* **34:** 275-280

Gordon, M.S. (1982) *Animal physiology: principles and adaptations.* 4th edition. MacMillan Publishing, New York, 68

Gordon, S.H. (1994) Effects of daylength and increasing daylength programmes on broiler welfare and performance. *World's Poultry Science Journal* **50:** 269-282

Gowe, R.S. and Fairfull, R.W. (1995) Breeding for resistance to heat stress. In: *Poultry production in hot climates.* Edit. Daghir, N.J., CAB International, Wallingford, 11-29

Haywood, M.P.S. (1990) Unpublished ADAS data

Hill, J.A. (1977) *The relationship between food and water intake in the laying hen.* PhD thesis. Huddersfield Polytechnic

Kettlewell, P.J. and Moran, P. (1990) A revised computer program for modelling heat stress in crated broiler chickens. Divisional Note DN1560, AFRC Institute of Engineering Research, Silsoe

Knight, C.D., Wuelling, C.W., Atwell, C.A. and Dibner, J.J. (1994) Effect of intermittent periods of high environmental temperature on broiler performance responses to sources of methionine activity. *Poultry Science* **73:** 627-639

Kutlu, H.R. and Forbes, J.M. (1994) Self selection for ascorbic acid by broiler chicks in response to changing environmental temperature. *British Poultry Science***:** **35** 820-821

Longhouse, A.D., Ota, H., Emerson, R.E., Heishman, J.O. (1968) Heat and moisture design data for broiler houses. *Transactions of the American Society of Agricultural Engineers* **11:** 694-700

Lott, B.D., May, J.D. and Simmons, J.D. (1998) Effective air temperature for broilers. *Poultry Science* **77:** Supplement, 4

MacLeod, M.G., Tullett, S.G. and Jewitt, T.R. (1980) Circadian variation in the metabolic rate of growing chickens and laying hens of a broiler strain. *British Poultry Science* **21:** 155-159

McKee, J.S., Harrison, P.C. and Riskowski, G.L. (1997) Effects of supplemental ascorbic acid on the energy conversion of broiler chicks during heat stress and feed withdrawal. *Poultry Science* **76:** 1278-1286

Mahmoud, K.Z. and Edens, F.W. (1998) Influence of vitamin C on heat shock protein 70 expression in five-week-old white leghorn chickens subjected to acute heat stress. *Poultry Science* **77:** Supplement, 77

Maloney, S.K. (1999) Heat storage, not sensible heat loss, increases in high temperature, high humidity conditions. *World's Poultry Science Journal* **54:** 347-352

May, J.D., Lott, B.D. and Simmons, J.D. (1997) Water consumption by broilers in high cyclic temperatures: bell versus nipple waterers. *Poultry Science* **76:** 944-947

Ministry of Agriculture, Fisheries and Food (1993) *Heat stress in poultry.* Booklet PB 1315. MAFF Publications, London

Mitchell, M.A. (1985) Effects of air velocity on convective and radiant heat transfer from domestic fowls at environmental temperatures of 20° and 30°C. *British Poultry Science* **26:** 413-423

Mitchell, M.A. and Hunter, R.R. (1996) Effects of chronic heat stress upon intestinal absorbtion of DL-methionine and methionine hydroxy analogue *in vivo* in the broiler chicken. *World's Poultry Science Association (UK Branch), Proceedings of spring meeting,* Scarborough, 24-25

Mitchell, M.A. and Sandercock, D.A. (1995) Creatine kinase isoenzyme profiles in the plasma of the domestic fowl (*Gallus domesticus*): effects of acute heat stress. *Research in Veterinary Science* **59:** 30-34

Monteith, J.L. (1974) Specification of the environment for thermal physiology. In: *Heat loss in animals and man*. Edit. Monteith, J.L. and Mount, L.E. Butterworths, London, 1-18

Mount, L.E. (1974) The concept of thermal neutrality. In: *Heat loss in animals and man.* Edit. Monteith, J.L. and Mount, L.E., Butterworths, London, 425-440

Okan, F., Kutlu, H.R., Baykal, L. and Canogullari, S. (1996a) Effect of wet feeding on laying performance of Japanese quail maintained under high environmental temperature. *World's Poultry Science Association (UK Branch), Proceedings of spring meeting.* Scarborough, 123-124

Okan, F., Kutlu, H.R., Canogullari, S. and Baycal, L. (1996b) Influence of dietary supplemental ascorbic acid on laying performance of Japanese quail reared under high environmental temperature. *World's Poultry Science Association (UK Branch). Proceedings of spring meeting,* Scarborough, 125-126

Oke, T.R. (1978) *Boundary layer climates.* Methuen and Co., London

Owen, J.E. (1994) Structures and materials. In: *Livestock housing* Edit. Wathes, C.M. and Charles, D.R., CAB International, 183-248

Ozcan, Z. and Kutlu, H.R. (1997) Effect of wet feeding and dietary supplemental ascorbic acid on performance of heat stressed broiler chicks. *World's Poultry Science Association (UK Branch). Proceedings of spring meeting,* Scarborough, 60-61

Reece, F.N., Deaton, J.W. and Kubena, L.F. (1971) Effects of high temperature and humidity on heat prostration of broiler chickens. *American Society of Agricultural Engineers, Winter meeting, 1971*

Richards, S.A. (1974) Aspects of physical thermoregulation in the fowl. In: *Heat loss in animals and man.* Edit. Monteith, J.L. and Mount, L.E., Butterworths, London, 255-276

Richards, S.A. (1976) Evaporation water loss in domestic fowls and its partition in relation to ambient temperature. *Journal of Agricultural Science, Cambridge,* **87:** 527-532

Richards, S.A. (1977) The influence of loss of plumage on temperature regulation in laying hens. *Journal of Agricultural Science, Cambridge,* **87:** 393-398

Romijn, C. and Lokhorst, M. (1966) Heat regulation and energy metabolism in the domestic fowl. In: *Physiology of the domestic fowl.* Edit. Horton-Smith, C. and Amoroso, E.C., Oliver and Boyd, Edinburgh, 211-217

Runge, G. (1991) Which fan where. In: *Shed environment project newsletter.* September 1991. Queensland Department of Primary Industries

Sainsbury, D.W.B. (1959) Electricity and problems of animal environmental control. *Electricity Development Association. Publication No.1897*

Saville, C.A., Clark, J.A. and Charles, D.R. (1978) House heat balance. *World's Poultry Science Association (UK Branch) Spring meeting*

Sinurat, A.P. and Balnave, D. (1986) Free-choice feeding of broilers at high temperatures. *British Poultry Science* **27:** 577-584

Smith, A.K. (1981) Poultry housing problems in the tropics and subtropics. In: *Environmental aspects of housing for animal production.* Edit. Clark, J.A., Butterworths, London

Sutcliffe, N.A., King, A.W.M. and Charles, D.R. (1987) Monitoring poultry house environment. In: *Computer applications in agricultural environments.* Edit.Clark, J.A., Gregson, K. and Saffell, R.A., Butterworths, London, 207-218

Sturkie, P.D. (1976) *Avian physiology.* Springer-Verlag, New York

Teeter, R.G and Belay, T. (1996) Broiler management during heat stress. *Animal Feed Science Technology* **58:** 127-142

Teeter, R.G., Wiernusz, C.J. and Belay, T. (1996) Animal nutrition in the 21st century. A poultry perspective. *Animal Feed Science Technology* **58:** 37-47

Tucker, S.A. and Walker, A.W. (1992) Hock burn in broilers. In: *Recent advances in animal nutrition.* Edit. Garnsworthy, P.C., Haresign, W. and Cole, D.J.A., Butterworth Heineman, Oxford, 33-50

Walker, A.W. (1995) Feeding regimes for broilers. *ADAS Poultry Progress,* November 1995, 4-6

Wathes, C.M. (1978) *Sensible heat transfer from the fowl.* PhD thesis, University of Nottingham

Wathes, C.M. (1994) Air and surface hygiene. In: *Livestock housing.* Edit. Wathes, C.M. and Charles, D.R., CAB International, Wallingford, 123-148

Wathes, C.M. and Clark, J.A. (1981) Sensible heat transfer from the fowl: radiative and convective heat losses from a flock of broiler chickens. *British Poultry Science* **22:** 185-196

Wathes, C.M, Gill, B.D. and Back, H.L. (1981) The effects of temperature on broilers: a simulation model of the responses to temperature. *British Poultry Science* **22:** 483-492

Wilson, E.A. (1986) *Surface energy exchange in poultry houses.* BSc thesis, University of Nottingham

Yahav, S. (2000) Relative humidity at moderate ambient temperatures: its effect on male broiler chcikens and turkeys. *British Poultry Science* **41:** 94-100

Yahav, S., Goldfeld, I., Plavnik, I. and Hurwitz, S. (1994) Physiological responses of broiler chickens and turkeys at high ambient temperature to relative humidity. Proceedings *World's Poultry Science Association 9th European Poultry Conference*, Glasgow, Volume I, 133-134

Yalcin, S., Settar, S., Ozkan, S. and Cahaner, A. (1997a) Comparative evaluation of three commercial broiler stocks in hot versus temperate climates. *Poultry Science* **76:** 921-929

Yalcin, S., Testik, S., Ozkan, S., Settar, P, Celen, F. and Cahaner, A. (1997b) Performance of naked neck and normal broilers in hot, warm, and temperate climates. *Poultry Science* **76:** 930-933

Zhou, W.T., Fujita, M., Ito, T. and Yamamoto, S. (1997) Effects of early heat exposure on thermoregulatory responses and blood viscosity of broilers prior to marketing. *British Poultry Science* **38:** 301-306

Zhou, W.T., Fujita, M., Yamamoto, S., Iwisaki, K., Ikawa, R., Oyama, H. and Horikawa, H. (1998) Effects of glucose in drinking water on the changes in whole blood viscosity and plasma osmolality of broiler chickens during high temperature exposure. *Poultry Science* **77:** 644-647

4

LIGHTING PROGRAMMES FOR LAYING HENS

D.R. Charles, P.D. Lewis and the late S.A. Tucker

The problems

Users are confronted with the need to choose photoperiodic regime, light intensity, light colour and light source. A great deal of experimental information has been generated and therefore summaries of practical recommendations have frequently been attempted, (e.g. Tucker and Charles, 1992; Charles and Tucker, 1992). This review provides more detail, with references to the supporting experimental evidence.

Response to photoperiod

The use of supplementary lighting to stimulate egg production was researched as early as the 1920s and 1930s, (e.g. Brown, 1925, quoted by Morris, 1999; Parkhurst, 1928; Rhys and Parkhurst, 1931). Wetham (1933) realised that the operative factor controlling ovulation was the seasonal change in daylength and not the daylength *per se*. At first it was sometimes argued that the stimulation of winter production with supplementary lighting merely redistributed the egg production within the year, but it was later realised that this was not so (Morris, 1968). The foundations of modern understanding of the subject and the basis of most modern practice probably date from the 1950s and 1960s (e.g. Morris and Fox, 1958 a and b; King, 1959 and 1961). Morris (1968) comprehensively reviewed the literature to that date and offered key principles of application, most of which still stand. Morris (1994 and 1999) provided updated reviews and recommendations.

A general principle of the photoperiodism of the chicken is that increasing daylength advances sexual maturity in the immature bird, and tends to stimulate ovulation after maturity. Decreasing photoperiods have the opposite effect (Morris, 1994). Early sexual maturity results in a higher number of eggs to a fixed finishing age, but smaller eggs (Morris, 1999).

For many years the standard practice in windowless houses in the UK was rearing on a short day of 6 to 8 hours, followed by a step-up in daylength at the rate of about 15 to 20 minutes per week from point of lay, reaching a maximum of 17 hours. Most debate, and recent experimental work, has concerned the age of introducing the step-up and the rate of step-up (see below). However Morris (1968) pointed out that there are probably many systems of lighting capable of maximising economic performance. To encourage feeding and drinking it is usual to start day old chicks on long days, but these must be discontinued by about two to three weeks of age in short day rearing regimes.

For applications such as free range or houses with windows, where shorter days than the prevailing natural daylength cannot be applied, programmes using supplementation of natural light by artificial light have usually been recommended (based on Morris, 1968). During rearing, a gradual step-down in daylength is used, from long days at day old, reaching the prevailing natural daylength at point of lay. The exact age at point of lay is often recommended by the breeding company. In the laying phase daylength is stepped up at the rate of about 20 minutes per week, starting at the natural daylength occurring at point of lay. Once daylength reaches 17 hours it is held at that length by supplementary lighting, though an EU directive and forthcoming UK codes of practice for animal welfare may dictate a 16 hour maximum daylength. It is important to avoid a reduction in daylength during lay.

At the time of writing there was much discussion of potential complications for the photoperiodism of organic flocks (Gordon and Charles, 2002). This was due to the requirement by the organic standards of a maximum of 16 hours light per day.

Shanawany and Morris (1980) confirmed that the principles discovered earlier still applied to the breeds of the late 1970s, though Shanawany (1983) found indications of breed differences in response. By that time it was suspected that modern breeds could be brought into lay earlier.

Charles and Tucker (1993) applied four light patterns, designed to give a range of ages at 50% lay, to four breeds. Despite significant effects on age at 50% rate of lay, light pattern had no significant effect on any performance trait, as measured from 20 to 80 weeks of age. They concluded that modern hybrids were so genetically prolific that they were predisposed to ovulate and were becoming refractory to lighting treatments formerly regarded as influential. The suggestion current in the industry at the time that rapid early increments in daylength may be used for some modern hybrids without compromising performance was supported. However Tucker and Charles (1992) saw no reason to change the practice, current at the time, of rearing on short days and then giving a 1 hour increment at 18 weeks, a further 1 hour at 19 weeks, and weekly increments of 15 minutes thereafter until 17 hours was reached. They noted that a range of ages for the initial increment was probably acceptable, provided that the birds reached their target body weight as recommended by the breeding company.

Lewis *et al.* (1996b) provided information on the achievement of desired ages at maturity. Birds reared on 10 hours light per day matured 8 days earlier than birds reared on 8 hours light per day. A 5 hour increment from 8 to 13 hours light at 84 days of age advanced maturity by 23 days compared with a constant 8 hour day, but only by 6 days when applied at 119 days of age. A reduction from 13 to 8 hours at 84 days of age delayed maturity by 22 days. Sensitivity to daylength changes varied with age and daylength change was more important than the absolute daylength. Lewis *et al.* (1997), in work in which the photoperiod was stepped up from 8 hours to 8, 10, 13 or 16 hours at 42, 63, 84, 105, 126 or 142 days of age, fitted functions for the effects of photoperiod and of age at step-up on performance factors. They concluded that the step-up was most effective in advancing age at first egg if given between 63 and 84 days. There were genetic differences in photosensitivity. It was the body weight at first egg rather than age which influenced egg weight (Lewis *et al.*, 1994).

Lewis and Morris (1998b) analysed a large number of publications on photostimulation and maturity. They concluded that age at first egg, body weight at first egg and egg weight, among other factors, increased linearly with age at photostimulation.

Intermittent lighting programmes

Rowland (1985) reviewed 75 papers on intermittent regimes. Since then there have been several important findings, some of which are as follows.

Morris *et al.* (1988) found a feed saving of approximately 2%, (though not quite significant), using the Cornell system of 2L:4D:8L:10D, without depressing egg output, compared with a step-up programme, though there was a reduction in egg size if the regime was applied abruptly at 18 weeks of age to pullets which had been reared on short days. Application from 21 or 24 weeks gave the same egg numbers and egg size as a step-up programme.

Midgley *et al.* (1988) explored the application for typical European brown-egg stocks of the *Bio-mittent*® programme, involving 15(0.25L:0.75D):0.25L:0.5D:0.25L:8D introduced at 37 weeks of age. (*Bio-mittent* is a registered trade mark of the Ralston Purina Company, St. Louis, Missouri,USA). Mean feed intake was 3.8% lower than the control between 18 and 72 weeks. Rate of lay and egg weight were similar to the control, provided that protein intake was maintained. Morris *et al.* (1990) concluded that the *Bio-mittent* regime can be introduced at point of lay. However note that the UK code of practice for animal welfare (MAFF, 1987) requires at least 8 hours of lighting per day, and the *Bio-mittent* regime only provides 4.25 hours per day.

Tucker and Charles (1993) found that a short cycle 4(3L:3D) regime, introduced at 20 weeks of age, was associated with a feed intake saving which varied across breed and rearing treatment between 1% and 4%. Egg weight and shell thickness were increased, but rate of lay and egg output were decreased by intermittent lighting compared with the control. Field experience has shown that introducing the regime later in lay allows it to be used without the depression in egg numbers, but that remains to be investigated scientifically.

Morris and Butler (1995) described a pattern using 24(0.25L:0.75D), designed to combine the increased egg size and shell thickness associated with symmetrical intermittent programmes with the reduction in feed intake associated with programmes reducing activity. This pattern (the Reading system) gave 2% fewer eggs than step up programmes, but increased egg size and shell thickness, and gave 6% lower feed intake than step-up. However it contravenes the UK codes of practice since it only provides 6 hours of illumination per day.

Lewis *et al.* (1992), in a review of 36 reports, suggested that intermittent lighting generally reduces mortality. Lewis *et al.* (1996a) calculated in a further review that mortality tended to increase as the amount of illumination increased.

Note that since intermittent regimes tend to reduce feed intake, probably by virtue of reduced activity and therefore energy requirement, it is important to ensure that the intake of the non-energy nutrients is maintained if intermittent schemes are used. Note also that feed savings due to intermittent lighting are likely to be less when applied at the correct temperature and with efficient feeding systems and good feather cover. Midgley *et al.* (1988) found that when intermittent lighting was compared with step up across temperatures and feeding systems, the savings due to lighting were less when feed intake was already low.

Since the various programmes give different sets of effects a summary table may be helpful.

Table 4.1 SUMMARY OF SOME EFFECTS OF INTERMITTENT LIGHTING TREATMENTS

Pattern	*Feed saving*	*Egg production*	*Egg weight*	*Shell thickness*	*Notes*
Cornell	Yes, but non-significant	Similar	Reduced; similar if applied later	Similar	
Bio-mittent	Yes	Similar	Similar	Similar	Contra-venes 1987 UK welfare codes
Short cycle	Yes	Reduced if introduced early in lay	Increased	Increased	Contra-venes EU directive
Reading	Yes	Reduced	Increased	Increased	Contra-venes EU directive and UK welfare codes

Ahemeral programmes

Ahemeral light programmes are defined as those in which the total hours of light and darkness add up to other than 24 hours. Discussion of their occasional use to improve egg weight and shell thickness, but at the expense of egg numbers if applied early in lay, has occurred since their development in the late 1960s and early 1970s. Foster (1968) compared 23, 24 and 25 hour cycles and proposed that individual birds have a "natural rhythm". Morris (1973) suggested that a suitable ahemeral daylength might be 28 hours, since this gives 6 long-days in a 7-day solar week. Note, however, that ahemeral light regimes contravene EU directives.

Morris and Bhatti (1978) found that a contrast between bright and dim light could be used to set the ahemeral phasing while always providing light for working in the building, but this would contravene the UK codes of practice for animal welfare (MAFF, 1987), which call for "...a period of darkness in each 24 hour cycle, ...". Rose *et al.* (1985) found that the bright and dim phases need not be from the same source type if two circuits were required for ease of dimming. Shanawany (1982) reviewed the earlier literature on the responses of production traits to cycle length, and Shanawany (1990) reviewed the effects of ahemeral programmes on egg quality.

The effect can be switched on and off within a laying cycle. Morris (1978) recommended that when changing back to a 24 hour basis the difference between the two photoperiods should be calculated and the difference added to the prevailing light period in order to find the appropriate light period for the 24 hour cycle. Thus 9L:18D (i.e. 9 + 18 = 27 = 24 + 3) could be followed by 12L:12D, (where 12 = 9 + 3). But there are important differences between application early and late in lay, quantitatively reviewed by Shanawany (1992). Early in lay the reduction in rate of lay may be fairly pronounced, but he summarised work indicating that if 28 hour days are not introduced until 55 weeks of age, then the egg weight and shell thickness improvements can often be achieved without a reduction in rate of lay.

Shanawany *et al.* (1993) showed why such effects occur, in an experiment comparing birds ovulating in short clutches of 6 eggs or less with those ovulating in clutches of more than 6 eggs. Long-day ahemeral light gives a longer mean intra-sequence interval between ovulations, but reduces the number of pause days between clutches. If the "hen's natural rhythm", (Foster, 1968), at the time is long then egg production can increase.

Light intensity

Morris (1968) described the effect of light intensity in the range 0.12 to 25 lux on rate of lay. Egg production was depressed at very low intensities and reached a plateau value at about 10 lux, which was regarded as a suitable intensity for the next 25 years or so.

$$y = 232.4 + 15.18x - 4.256x^2$$

where y = egg yield and x = $\log_{10}$ light intensity (lux)

Yet some years later Hill *et al.*(1988) found no effect on egg output of intensities over the range 2 to 45 lux. Tucker and Charles (1993) found no consistent response over the range 0.75 to 12.4 lux. Morris (1994) suggested that about 5 lux is adequate. Thus the UK work led to the conclusion that it seems that modern hybrids are less sensitive to low levels of light intesity than were the earlier birds, though it is probably still wise to provide about 10 to 20 lux for the purposes of working conditions and inspecting the birds. Lewis *et al.* (1997) found evidence of genetic differences in photosensitivity. The UK codes of practice for animal welfare (MAFF, 1987) suggest that "Enough light should be available to enable all birds to be seen clearly when they are being inspected............".

Recent work in Canada found some evidence for a depression in ovary weight and egg production at 1 lux compared with higher intensities in an experiment using up to 500 lux (Renema *et al.,* 1998). The effects varied with breed across four breeds, including white and brown birds.

Lewis and Morris (1999) quantitatively reviewed publications on responses to light intensity and fitted response curves enabling economic optima to be calculated. They distinguished recent work from older work and took into account egg production, egg weight, mortality and feed intake. They concluded that the economic optimum is close to 5 lux, but they recognised the need to allow for working conditions and perceived bird welfare, and therefore recommended about 10 lux.

Light colour and source

Morris (1968) reported that it is generally assumed that the fowl responds mainly to the red and orange part of the visible spectrum, and it is therefore probably important that light sources used in laying houses have adequate red and orange. Many white light sources are therefore suitable. Lewis and Morris (2000) described the differences in photoreception between poultry and humans;

poultry being more sensitive to the blue and red parts of the spectrum and in addition they are sensitive to UV-A radiation. Growth and behavioural responses were said to depend principally on retinal photoreception, whereas photosexual responses are mainly to direct hypothalamic reception. Red light is more sexually stimulatory than blue or green, whereas growth, at least in broilers and turkeys, is inferior under red illumination (Lewis and Morris, 2000).

Hill *et al.* (1988) found a non-significant increase in feed intake attributable to fluorescent light when compared with tungsten filament lamps, which if repeatable would financially outweigh the saving in electricity cost. The reasons for this finding were not clear. Nuboer *et al.*(1992) concluded that chickens see fluorescent lamps as flickering, yet Widowski *et al.* (1992) found that given a choice hens preferred fluorescent light. Boshouwers and Nicaise (1992) found that broilers reacted differently to fluorescent lights of high and low frequency, perhaps because the high frequency were not perceived as flickering. The frequencies were 26 kHz and 100 Hz respectively. After some years of debate in the UK industry, Lewis and Morris (1998a) reviewed the literature and concluded that there was no evidence of any consistent detrimental effect of fluorescent lighting on reproductive performance.

The sensitivity of poultry to UV-A radiation could have significance for performance and behaviour, as suggested for turkeys by Lewis *et al.* (2000a), and for layer pullets by Lewis *et al.* (2000b).

Light proofing

For many years it has been considered good practice to light proof laying and rearing houses so that the intensity in the dark phase is below 0.4 lux. This practice has been based on the findings of Morris and Owen (1966) that the rate of lay at 0.4 lux was not significantly different from that in darkness.

Lewis *et al.* (1999) investigated the involvement of light intensity in sexual development and concluded that the threshold for an optimal response lies between 0.9 and 1.7 lux. However, when an 8 hour period of light at an illuminance of only 0.03 lux was added to a normal 8 hour photoperiod, sexual maturity was advanced by about a week compared with short-day controls. Whilst this advance may have been affected by the dim light producing a phase shift in the photo-inducible period, rather than the dim light itself being stimulatory, it did indicate that there is no known intensity of light that can be universally treated as darkness.

Some recommendations

- There is probably no reason for drastic changes in the conventional practice for windowless houses, of short-day rearing followed by a step-up in daylength to 17 hours of light per day, this daylength being maintained throughout lay. (Note that a forthcoming UK code of practice may limit the maximum hours of light per day to 16.) Modern hybrids may be tolerant of earlier introduction of the step-up, though care should be taken that the birds reach the target body weight, as recommended by the breeding company, prior to photostimulation.

- Supplementary lighting programmes can be used for windowed housing, or for free range. A gradual step-down in daylength is provided from long-days at day old, reaching the prevailing natural daylength at point of lay. Then a gradual step-up is used to 17 hours light per day, and this daylength is maintained throughout lay. (Note that a forthcoming UK code of practice may limit the maximum hours of light per day to 16).

- Modern hybrids may be more tolerant of light intensity than earlier birds, but a minimum of 10 lux should be used for operational and inspection reasons.

- The traditional recommendation that light proofing be such that the intensity during "lights off" is below 0.4 lux is still a practical recommendation for both the rearing and laying periods, but for complete control of sexual maturity light proofing needs to be absolute.

- The breeding companies have recommendations on lighting for their birds.

- Always ensure that lighting programmes conform to EU directives, regulations and codes of practice, which occasionally change.

References

Boshouwers, F.M.G. and Nicaise, E. (1992) Responses of broiler chickens to high-frequency and low-frequency fluorescent light. *British Poultry Science* **33:** 711-717

Charles, D.R. and Tucker, S.A. (1992) Lighting for egg production. *ADAS Poultry Progress* No. 17, November 1992

Charles,D.R. and Tucker, S.A. (1993) Resonses of modern hybrid laying stocks to changes in photoperiod. *British Poultry Science* **34:** 241-254

Foster, W.H. (1968) The effect of light-dark cycles of abnormal lengths upon egg production. *British Poultry Science* **9:** 273-274

Gordon, S.H. and Charles, D.R. (2002) *Niche and organic chicken products: their technology and scientific principles.* Nottingham University Press (In press)

Hill, J.A., Charles, D.R., Spechter, H.H., Bailey,R.A. and Ballantyne, A.J. (1988) Effects of multiple environmental and nutritional factors on laying hens. *British Poultry Science* **29:** 499-512

King, D.F. (1959) Artificial light for growing and laying birds. *Alabama Agricultural Experimental Station Progress Report* 72

King, D.F. (1961) Effects of increasing, decreasing and constant lighting treatments on growing pullets. *Poultry Science* **40:** 479-484

Lewis, P.D and Morris, T.R. (1998a) Responses of domestic fowl to various light sources. *World's Poultry Science Journal* **54:** 7-25

Lewis, P.D. and Morris, T.R. (1998b) A comparison of the effects of age at photostimulation on sexual maturity and egg production in domestic fowl, turkeys, partridges and quail. *World's Poultry Science Journal* **54:** 119-128

Lewis, P.D. and Morris, T.R. (1999) Light intensity and performance of domestic pullets. *World's Poultry Science Journal* **55:** 241-250

Lewis, P.D. and Morris, T.R. (2000) Poultry and coloured light. *World's Poultry Science Journal* **56:** 189-207

Lewis, P.D., Morris, T.R. and Perry, G.C. (1996a) Lighting and mortality rates in domestic fowl. *British Poultry Science* **37:** 295-300

Lewis, P.D., Perry, G.C. and Morris, T.R. (1996b) Effect of constant and changing photoperiods on age at first egg and related traits in pullets. *British Poultry Science* **37:** 885-894

Lewis, P.D., Morris, T.R. and Perry, C.G. (1999) Light intensity and age at first egg in pullets. *Poultry Science* **78:** 1227-1231

Lewis, P.D., Perry, G.C. and Morris, T.R. (1994) Effect of breed, age and body weight at sexual maturity on egg weight. *British Poultry Science* **35:** 181-182

Lewis, P.D., Perry, G.C. and Morris, T.R. (1997) Effect of size and timing of photoperiod increase on age at first egg and subsequent performance of two breeds of laying hen. *British Poultry Science* **38:** 142-150

Lewis, P.D., Perry, G.C., Sherwin, C.M. and Moinard, C. (2000a) Effect of ultraviolet radiation on the performance of intact male turkeys. *Poultry Science* **79:** 850-855

Lewis, P.D., Perry, G.C. and Morris, T.R. (2000b) Ultraviolet radiation and laying pullets. *British Poultry Science* **41:** 131-135

Lewis, P.D., Perry, G.C., Morris, T.R. and Midgley, M.M. (1992) Intermittent lighting regimes and mortality rates in laying hens. *World's Poultry Science Journal* **48:** 113-120

Midgley, M.M., Morris, T.R. and Butler, E.A. (1988) Experiments with the Bio-mittent lighting system for laying hens. *British Poultry Science* **29:** 333-342

Ministry of Agriculture, Fisheries and Food (1987) *Codes of recommendations for the welfare of livestock. Domestic fowls.* MAFF Publications, London

Morris, T.R. (1968) Light requirements of the fowl. In: *Environmental control in poultry production.* Edit. Carter, T.C., Oliver and Boyd, Edinburgh

Morris, T.R. (1973) The effects of ahemeral light and dark cycles on egg production in the fowl. *Poultry Science* **52:** 423-445

Morris, T.R. (1978) The photoperiodic effect of ahemeral light-dark cycles which entrain circadian rhythms. *British Poultry Science* **19:** 207-212

Morris, T.R. (1994) Lighting for layers - what we know and what we need to know. *World's Poultry Science Journal* **50:** 283-287

Morris, T.R. (1999) Sexual maturity, lighting and layer performance. Poultry lighting seminar, University of Bristol, 10-12

Morris, T.R. and Bhatti, B.M. (1978) Entrainment of oviposition in the fowl using bright and dim light cycles. *British Poultry Science* **19:** 341-348

Morris, T.R. and Butler, E.A. (1995) New intermittent lighting programme (the Reading System) for laying pullets. *British Poultry Science* **36:** 531-535

Morris, T.R., and Fox, S. (1958a) Light and sexual maturity in the domestic fowl. *Nature* **181:** 1453-1454

Morris, T.R. and Fox, S. (1958b) Artificial light and sexual maturity in the fowl. *Nature* **182:** 1522-1523

Morris, T.R., Midgley, M.M. and Butler, E.A. (1988) Experiments with the Cornell intermittent lighting system for laying hens. *British Poultry Science* **29:** 352-332

Morris, T.R., Midgley, M.M. and Butler, E.A. (1990) Effect of age at starting bio-mittent lighting on performance of laying hens. *British Poultry Science* **31:** 447-455

Morris, T.R. and Owen, V.M. (1966) The effect of light intensity on egg production. In: *Proceedings 13th World Poultry Congress*, Kiev, 458-461

Nuboer, J.F.W., Coemans, M.A.J.M. and Vos, J.J. (1992) Artificial lighting in poultry houses: do hens perceive the modulation of fluorescent lamps as flicker? *British Poultry Science* **33:** 123-133

Parkhurst, R.T. (1928) Artificial light for late hatched pullets. *Eggs.* Scientific Poultry Breeders Association. December 1928, 270-271

Renema, R.A., Robinson, F.E., Feddes, J.J.R., Zuidhof, M.J., Newcombe, M. and Kulenkamp, A. (1998) Effects of light intensity and strain on carcass traits ovarian morphology and egg production parameters in egg-type chickens. *Poultry Science* **77:** Supplement, 34

Rhys, I.W. and Parkhurst, R.T. (1931) Methods of artificially lighting winter layers. *Harper Adams Agricultural College Bulletin,* **6**

Rowland, K.W. (1985) Interrupted lighting patterns for laying chickens:a review *World's Poultry Science Journal* **41:** 5-19

Rose,S.P., Bell,M.and Michie,W. (1985) Comparison of artificial light sources and lighting programs for laying hens on long ahemeral light cycles. *British Poultry Science* **26:** 357-365

Shanawany, M.M. (1982) The effect of ahemeral light and dark cycles on the performance of laying hens - a review. *World's Poultry Science Journal* **38:** 120-126

Shanawany, M.M. (1983) Sexual maturity and subsequent laying performance of fowls under normal photoperiods -a review 1950-1975. *World's Poultry Science Journal* **39:** 38-46

Shanawany, M.M. (1990) Ahemeral light cycles and egg quality. *World's Poultry Science Journal* **46:** 101-108

Shanawany, M.M. (1992) Response of layers to ahemeral light cycles incorporating age at application and changes in effective photoperiod. *World's Poultry Science Journal* **48:** 156-164

Shanawany, M.M. and Morris, T.R. (1980) Light, sexual maturity and subsequent performance. *World's Poultry Science Association (UK Branch) Summer meeting.*

Shanawany, M.M., Morris, T.R., and Pirchner, F. (1993) Influence of sequential length on the response to ahemeral lighting late in lay. *British Poultry Science* **34:** 873-880

Tucker, S.A. and Charles, D.R. (1992) Lighting for laying hens. *ADAS Poultry Progress* No.16, May 1992

Tucker, S.A. and Charles, D.R. (1993) Light intensity, intermittent lighting and feeding regimen during rearing as affecting egg production and egg quality. *British Poultry Science* **34:** 255-266

Wetham, E.O. (1933) Factors modifying egg production with special reference to seasonal changes. *Journal of Agricultural Science* **23:** 383-419

Widowski, T.M., Keeling, L.J. and Duncan, I.J.H. (1992) The preferences of hens for compact fluorescent over incandescent lighting. *Canadian Journal of Animal Science* **72:** 203-211

FURTHER INFORMATION

Several items reported in this book are based on ADAS research and consultancy. For further information please consult the following:

ADAS Poultry R&D team:

Andrew Walker
ADAS Land Research Centre
Gleadthorpe
Meden Vale
Mansfield
Nottinghamshire
NG20 9PF

Phone 01623 844331
Fax 01623 844472
e-mail andrew.walker@adas.co.uk

INDEX